UNDER CONSTRUCTION. BUILDING THE MATERIAL AND THE IMAGINED WORLD

Eike-Christian Heine (ed.)

KULTUR UND TECHNIK

Schriftenreihe des Internationalen Zentrums
für Kultur- und Technikforschung (IZKT)
der Universität Stuttgart

Herausgegeben von:
*Reinhold Bauer, Helmut Bott, Franz Brümmer, Georg Maag,
Wolfram Pyta, Ortwin Renn, Werner Sobek, Antje Stokman*

BAND 30

LIT

UNDER CONSTRUCTION. BUILDING THE MATERIAL AND THE IMAGINED WORLD

Eike-Christian Heine (ed.)

LIT

BIBLIOGRAFISCHE INFORMATION
DER DEUTSCHEN BIBLIOTHEK

Die Deutsche Bibliothek verzeichnet diese Publikation in der Deutschen Nationalbibliografie; detaillierte bibliografische Daten sind im Internet über http://d-nb.de abrufbar.

ISBN 978-3-643-90700-4

Verlagskontakt:
Fresnostr. 2 D-48159 Münster
Tel.: 0251 62 03 20 Fax.: 0251 23 19 72
E-Mail: lit@lit-verlag.de http://www.lit-verlag.de

Satz: Nora Heinzelmann
Entwurf Grafik: Jörg Schwertfeger | www.joergschwertfeger.ch

TABLE OF CONTENTS

PART 5: REPRESENTATION UNDER CONSTRUCTION

PART 6: CONCEPTUALISING CONSTRUCTION SITES

INTRODUCTION

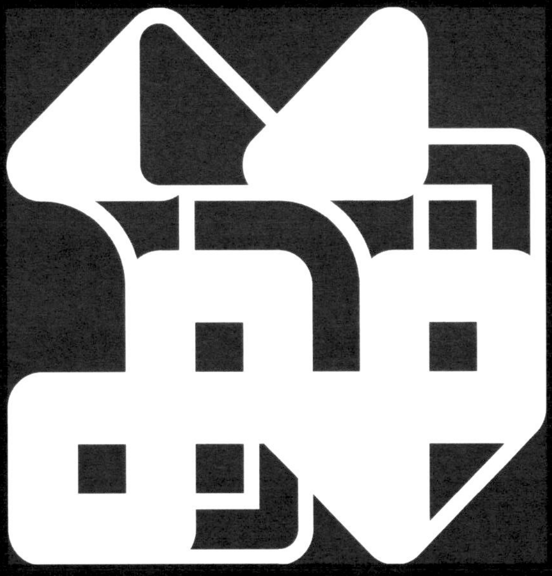

THE CRANES WERE EVERYWHERE. BUILDING THE MATERIAL AND THE IMAGINED WORLD

INTRODUCTION

Eike-Christian Heine

It seemed the cranes were everywhere. When the train gently rumbled into Stuttgart's terminus, the right hand site of the coach was not a cityscape, but a clutter of brown earth, of concrete not yet revetted, of machines swarming on the ground and of bright yellow truss reaching high into the blue August sky [fig. 1]. It came as no surprise that I found myself on a construction site. Indeed, I was expecting it. After a fierce debate that was covered nationwide, after years of protest and sometimes surprisingly violent clashes in the otherwise steady and even docile capital of Baden-Württemberg, a state-wide referendum nine months earlier had finally given the signal to completely rebuild the main station, to reroute the tracks in the heart of the city and to put rail traffic underground.

It was in 2012 when I moved to Stuttgart. When I settled in into the town new to me, I met people who for large parts no longer heatedly debated the question of the new station, but who told about the protests as if they had happened in another era. They reported about the discussions everywhere in the city, about the necessity to position yourself either against or in favour of the project known as Stuttgart 21, about the weekly protest marches on Monday, and about the plebiscite that – for most of the city – took the edge off

Fig. 1 The cranes reached high into the sky. Construction in Stuttgart's new *Europaviertel* city district, which is part of the highly controversial project Stuttgart 21 (2013).

the debate. Not that everyone is convinced. The protest is still alive, while the major works just a couple of meters away from the university's downtown campus progress slowly but steadily. And with this progress the number and location of the cranes change, but – like the protesters on Monday nights – they never disappeared so far.

It was amongst these cranes that a number of scholars came together. On two November days in 2013, 14 people from eight countries met in Stuttgart's new city library to discuss their research. The great white cube, 40 meters in each dimension and designed by the Korean architect Eun Young Yi, was opened in 2011 and was towering over its flat surroundings, where machines and workers were busy constructing the foundations of a new city block. It was a fitting place to meet, as we came to discuss the material and imagined worlds that were under construction at other places in time, where and when people were busy erecting their build environment.

There is one quite simple idea that forms the basis of our discussions and the articles that are collected here:[1] When people are building a house or a dam or practically any other thing, they are always making assumptions about the past, the present or the future. The intervention into the material world is accompanied by the construction and reconstruction of "worlds in the heads." By shifting attention away from the product and focussing on the production of architecture and infrastructure, the attention is shifted towards one of the central arguments of cultural studies. Instead of analysing the identity of objects, nations, architecture, science, technology or any other thing, the focus is turned towards the history of how these things were made by different people with their words, hands and tools. One of the important reference points of such a thinking certainly is Michel Foucault, who notes that the challenge of historical analysis is not "a theory of the knowing subject, but rather [...] a theory of discursive practise."[2]

One of the striking texts which illustrate that things imagined are often not less real than material objects is Benedict Anderson's study of nationalism. By writing about "imagined communities", he writes about language, literature, education, geography, archaeology and many other very real things, which, taken together, help to understand how the concept of the nation came to be. "Imagined" does not doubt that nations are powerful political ideas and entities, only that they are based on shared belief and on shared practice, not on some sort of natural or divine world order.

1 For different reasons some of the participants of the conference cannot be part of this volume. As a way of thanking them for their visit to Stuttgart, I would like to mention some of their work. Catarina Caetano da Rosa currently works on a transfer-history of Lisbon and Rio de Janeiro. Her dissertation is about a completely different topic: *Operationsroboter in Aktion. Kontroverse Innovationen in der Medizintechnik*, Bielefeld 2013. Francesca Russello Ammon gave a presentation about urban reconstruction; one of her articles is "Unearthing Benny the Bulldozer. The Culture of Clearance in Postwar Children's Books, in: *Technology and Culture*, vol. 53, 2012, 306-336. Eva Cancik Kirschbaum proofed not only to be a distinguished archaeologist and historian; she also proofed to be a highly entertaining keynote speaker. Probably closest to what she presented in Stuttgart is the following book: *Babylon. Wissenskultur in Orient und Okzident*, Berlin 2011 (with Ess, Margarete van and Marzahn, Joachim). Birte Förster gave a presentation about the Volta river project in Ghana and the power relations materialised in this infrastructure. From her publications see e.g.: "Wasserinfrastrukturen und Macht. Politisch-soziale Dimensionen technischer Systeme", in: Förster, Birte/Martin Bauch (ed.), *Wasserinfrastrukturen und Macht von der Antike bis zur Gegenwart*, Special Issue of *Historischen Zeitschrift* 63, Berlin 2014, 9–21.

2 Foucault, Michel, *The Order of Things*, New York 1973, xiv.

The discourses and practices have not always been there, but appeared somewhere in time and space, changed and even ended.

The notion that some things are imagined does not come out of innocent acceptance. The exact opposite is the case; it comes with a deep bewilderment of how things are, but also with critique and even revolt. Foucault analysis of the historical apriori or the order of things is always connected with his notion that some rules of the game should be changed; Benedict Anderson was exiled from Indonesia because of his critique of the military junta; and Stuttgart's Monday night demonstrators do not agree with the material reconstruction of parts of their city. Among other things they also fear that building tunnels will bring underground layers of gypsum in contact with groundwater, and that the following reactions would start to move the ground in large scale, with uncontrollable results for the city above. The material and imagined worlds are rarely built in agreement, often they tell of political power contested or of defiant natural elements.

Several studies already approach this tandem of the material and the imagined. Christine Wall for example studies how scarcities during and after the Second World War put traditional building methods in Britain under pressure. The implementation of innovation changed the characteristic of construction work. She especially stresses how an industry that mainly relied on the output of a worker (e.g. bricks laid during a day) became one where "the social construction of 'skill' now predominates."[3] This change is one amongst many different changes in practices, institutions and expectations of individuals. But she also believes that the conflict between capital and labour must not stay stuck in the old world of piece work, rather labour conflicts should be about "education and training". If the old ways continue, she argues, the hierarchical British class society will not change.[4]

Adrian Forty argues that a special construction material was part of utopian thinking. In *Concrete and Culture* he notes that writers for a long time had "imagined [...] a perfect, cement-based building material that would transform people's lives."[5] Tom Peters concentrates on the evolution of construction in great detail. His book *Building the nineteenth century* is an astonishing collection of techniques and innovations in this vital sector of the economy. Although focussing on technologies and practices of construction, he nonetheless argues that a new thinking has emerged alongside technological innovation. He claims that "building has helped form, and not merely followed, our cultural development."[6] It did so, amongst other things, by offering a field of activity where scientific knowledge and practical experience formed a successful relationship.[7]

Judith Schueler offers yet another context in which construction sites form material and ideological worlds and world views. Her research is part of

3 Wall, Christine, *An Architecture of parts. Architects, building workers and industrialisation in Britain 1940-1970*, London/New York 2013, 13.

4 Ibid., 14.

5 Forty, Adrian, *Concrete and culture. A material history*, London 2012, 8.

6 Peters, Tom Frank, *Building the nineteenth century*, Cambridge (Mass.)/London 1996, 356.

7 Ibid., 347.

the "Tensions of Europe" network that analyses technology and the making of Europe.[8] She studies the construction of the Gotthard railway tunnel, opened in 1880, as the "co-construction of technology and identity."[9] The building of infrastructure became a powerful symbol for an evolving Swiss nationalism. A similar case finds Chandra Mukerji in France during the reign of Louis XIV. What happened here was a "shift of attention to the built environment as a site for political action. Building and war", she continues, "were particularly potent tools for constructing a territorial state"[10]. Construction work for her necessarily is the construction of meaning, too:

> Humans are generally not content simply to admire and reflect upon stones, trees, water, birds, insects, and clay without moving them around. We end up constructing a humanly modified world in which we live [...] [and have the, ECH] tendency to move things around to give the world aesthetically or economically a satisfying form.[11]

Such positions form the context of the papers collected here. Ranging from the Early Modern Period and reaching until our present time, most authors in this volume concentrate on the nineteenth and twentieth century. Although the focus mainly is on construction projects in Europe, Istanbul, New York and Central Asia are also represented. The outcome is a broad temporal and spatial perspective on the material and imagined worlds that were built on and around construction projects.

Construction sites can become keys to understanding the constant change that characterised world cities in the twentieth century. Patrick Leitner shows how the chaotic and un-planned destruction and reconstruction of New York let to excessive amounts of noise and dirt. But he also shows that a vision of New New York gave the city's inhabitants utopian images that let them endure the hardships of everyday life. Ipek Şen and Şebnem Şoher tell the story of Istanbul becoming a permanent construction site and how Europe and the USA served as role-models. Their text is a warning to continue urban reconstruction without democratic moderation, and – although not mentioned expressively – the Taxim Square protests of 2013 give a telling example of the friction, construction projects can represent for whole societies. There is an unresolved question that develops out of Istanbul's continuous reconstruction as well as out of Stuttgart's permanent protest: To what degree has the narrative of utopian progress lost the convincibility today, it once held for New York in the early twentieth century?

While the story of New New York was a powerful narrative to understand and legitimate large scale reconstruction of the city, its character was different from construction sites in socialist utopias. Marie Collier shows how photography and photomontage of large construction sites became

8 The network's agenda and several relevant publications can be found at http://www.tensionsofeurope.eu/ (accessed 18-6-15).

9 Schueler, Judith, *Materialising Identity. The Co-construction of the Gotthard Railway a Swiss national Identity*, Amsterdam 2008, 139.

10 Mukerji, Chandra, *Territorial ambitions and the gardens of Versailles*, Cambridge 1997, 1.

11 Ibid., 304.

a key element of Soviet propaganda during the first Five Year Plan. The staging of the construction site as a metaphor for the ongoing construction of socialism is the *leitmotiv* that Marius Böttcher discusses as well. His analysis is characterised by the invasion of elements that were suppressed by construction activity and socialist propaganda in East Germany and that come sneaking back into life through the arts. Drawing from Freudian psycho-analysis, Marius work on construction sites in GDR cinematography insists that the material reality and changes in that material reality can have powerful impacts on individuals and whole societies.

A close interrelation between the nation and construction is already formulated in East Germany's national anthem (*Auferstanden aus Ruinen*, "Risen from ruins"). Such connections explore the articles of Liliana Iuga and Fillippo Menga in different times and places. Liliana shows how the construction of a church between the World Wars intersects in complex ways with questions of nationality, language and religion. In the Transylvanian city of Cluj, nationality and social class were ordered along religious lines. When the region became part of Rumania, the national government was lobbied to build a representative orthodox cathedral to leave a mark on the city dominated by German and Hungarian. She also shows how the practices of construction could tap the existing multi-ethnic traditions. The focus on construction work thus offers perspectives that could easily be overlooked if only architecture and the cityscape were analysed. It is my understanding that there also lies a reservoir of hope in that story for a multi-ethnic and multi-national Europe, even if it does not report of harmony and only of grudging cooperation. The latter is undoubtedly better than war, or the politics of nationalisation and ethnic cleansings, horrors from which Europe suffered at great lengths in the twentieth century. Fillippo Menga certainly shares such hopes for Central Asia, where the construction of a large dam in Tajikistan produces tensions. The construction project is part of a national political strategy that identifies infrastructures as carriers of economic development and as a symbolic source for national coherence. At the same time this policy seems to accept violent conflict with the nation's neighbours. While in Soviet time Moscow's power moderated between the Socialist Republics, war might be the outcome if not a path of political cooperation and moderation is taken today.

That politics and construction projects are often dependent on each other can also be seen in colonial spaces. While the construction of railroads in the Ottoman Empire, in the American West or in the Caribbean as well as the digging of the Suez and Panama Canals are well known example for "imperial infrastructures"[12], Sören Fischer takes the volume's question into Early Modern Times. The changing patterns of world trade, technological innovation and political power forced the rich nobility of Venetia away from the sea and into the city's hinterland. A large-scale colonisation project was started, during which canals, villages and villas were built. By closely looking at the representation of this newly build nature, Sören shows which

12 Laak, Dirk van, *Imperiale Infrastruktur. Deutsche Planungen für eine Erschließung Afrikas 1880 bis 1960*, Paderborn 2004.

imagined and material worlds were constructed simultaneously during the colonisation of the Terraferma.

Clemens Voigts writes about a different construction project in Early Modern Times, the erection of a large monolith in front of St. Peter's Cathedral. He describes how the technical process of moving an obelisk was staged in front of a huge crowd, how the leading engineer was able to successfully promote the occasion for his professional reputation and how for centuries to come a tradition formed that relied on both the engineering solution and the public staging of the construction site. My own article also tells of the technical and organisational challenges of a construction site and its representation in the public sphere. The Kiel Canal was built late in the nineteenth century and it until today allows passage between the Baltic and the North Sea for large ships. The engineers employed masses of workers and machines to form a stable canal out of defiant swamps and sands. During nine years of construction, a conservative public imagined its hope for a harmonic class society realised and was also building its understanding of the machine age.

Christoph Rauhut's study again takes a different perspective. He asks what happens to large narratives of modernity – the stories of rationalisation and mechanisation – if you take a very detailed look at a construction site. Christoph is arguing not to over-interpret building sites that at their time were staged as symbols of modernity. Historiography instead needs to take a closer look at "every-day" projects and he closely researches such a construction site in Switzerland. His findings reflect on the intellectual worlds we are constructing in our academic work.

Out of the many topics we debated in Stuttgart and that are collected in this volume, I would like to pick four that seem especially important: construction sites hold the chance to identify different forms of progress and to historicise these various progresses; on building sites social hierarchy and power were constructed and reconstructed; the building site holds the opportunity to identify counter-narratives that are overlooked if one only sees the product and not also the production-process of architecture and infrastructure; and finally one has to reflect a constitutive feature of construction sites: they are only ephemeral, as their existence is (mostly) only temporarily and as their spatial borders are very permeable to rubble going out, to construction material coming in and open to works of definition and reinterpretation.

The construction site is in this volume the key to understand the material, discursive and performative building of worlds and world views. One of the themes that appears repeatedly is progress. For the German historian Reinhart Kosselleck, progress is a central concept of modernity. From around 1800, political, scientific and technological revolutions happened and the pace of historical change accelerated significantly. In his understanding, progress became a narrative to rationalise and deal with the continuous and enormous transformation. With "progress" contemporaries came to terms with the situation of rapid technological, social and political change.[13] This is of course not only a passive reaction, the expectance of

13 Koselleck, Reinhart, *Vergangene Zukunft. Zur Semantik geschichtlicher Zeiten*, Frankfurt a. M. 1979, e.g. 59.

ongoing revolutions also shaped behaviour and thought, thus stabilising and accelerating transformation.

The articles show a broad spectrum of historical change: The volume tells of the utopian progress that, as a shared belief of a whole society, let people endure the unplanned and unchecked reconstruction of New York; large steam driven construction machinery was contesting the understanding of humanity's control over the elements; progress became a narrative to make sense of the nation's and man's growing power over nature; and there appears the conviction that technical culture is a culture of progress itself. Socialist progress took on similar forms to the case of New New York or Istanbul, of course with the significant difference that it was part of a rigorous state driven propaganda. Yet, while the narrative of social progress collapsed in Socialism, it remains uncertain what happens in capitalist and democratic societies today. Clearly for most of us to see, the fantastic promises of progress were not realised in the West (just compare the visions of New New York with the realities of today). One point in time where the story of progress collapsed was during the ecological revolution of the 1970's, which put a political influential doubt to the story of human's control of nature through technology. As far as this volume goes, the story of the construction under such new paradigms remains open (one might take a look at protests against the construction of new atomic power plants to show how the construction of the material world and the world in the heads changed). However, a historiography of technical culture needs to historicise progress.[14]

Construction sites were spaces where social hierarchies were produced and reproduced continuously. In the Early Modern period the technical expert became a common and well-known actor. Domenico Fontana was able to promote himself as an engineering star. He did so by staging the construction site as a public event and by opulently publishing about "his" moving of obelisks. Looking back at the conference and the volume I am surprised that the question of the architect and of authorship is not debated in other articles. The construction site seems the right place to reflect on star engineers and architects by evaluating their achievement in the production process of their respective time without reproducing the hyped categories of marketing that since Fontana many experts necessarily played so well.

While this remains as an open question, the construction site in this volume appears as a place where social hierarchy is reproduced. The nobility that once ruled over Europe's shipping lines to the East started to install themselves as rulers over a newly colonised Terraferma. Sören shows how the architectural staging of roads, canals and villas materialised the claim

14 After meeting in Stuttgart, Christoph and I decided to further approach the question of the construction site as a place of mixed temporality. We organised a session at the 2014 conference of the EAHN entitled: "Producing non-simultaneity: Construction sites as places of Progressiveness and Continuity." The papers are published in: Rosso, Michela, *Investigating and writing architectural history. Subjects, methodologies and frontiers. Papers from the Third EAHN International Meeting*, Torino 2014, http://www.eahn2014.polito.it/EAHN2014abstracts.pdf (accessed 17-6-15) and include: Cobb, Elvan, "Mixing Time: Ancient-Modern Intersections along the Western Anatolian Railways", 336-346; Rosenberg, Frida, "Steel as Medium. Constructing WGC, a Tallish Building in Postwar Sweden", 347-354; Vidal, Marisol, "Between Technological Effectiveness and Artisanal Inventiveness; Concreting Torres Blancas (1964-69)", 355-365; Kozlovsky, Roy, "The Glocal Construction Site and the Labour of Complex Geometry", 366-374.

of the Venetian nobility to rule over an imagined Eden and the people that populated it. The case of the Kiel Canal late in the nineteenth hundreds shows a different way in which construction sites can become places of social policy. The vision of a harmonic class society was formulated down to the layout of the barracks. Daniel Miller has convincingly argued what happened here: "the study of material culture often becomes an effective way to understand power, not as some abstraction, but as the mode by which certain forms of people become realised, often at the expense of others."[15]

But what about these others? What did the workers become when they were placed in barracks? And what about progress; what happened to nature, things, individuals and collectives when progress became reality? Both the idea of progress and the structures of social stratification, which were often imagined and materialised on the construction site, call for checks with – what I call for the sake of the argument – different historical realities. One way to do such necessary checks is to strive for the inclusions of the visions and interpretations of marginalised actors into history. With the words of Berthold Brecht it is to ask who built Thebes of seven gates: "In the books you will find the name of kings. Did the kings haul up the lumps of rock?"[16] Of course the kings only did the ritual shovelling that symbolises the beginning of construction. But the story of the workers is difficult to tell, as are so many histories of marginalised people.[17] In the case of the Kiel Canal, the imaginations of a harmonic society that came with the realisation of social policy proofed to produce hate and disgust for the barracks amongst the workers. The construction of the Kiel Canal also shows that nature has a material life of its own that can escape the understanding of the expert. There is a defiant side to materiality that can help us to better understand.

The volume's first section "Constructing Socialism" shows how building activity and communist propaganda worked hand in hand. It also illustrates how the experience of material reality of every-day socialism contested such imaginary worlds. The material here has the potential to upset aspirations of control of nature and society. For another political system, Liliana Iuga shows how the inclusion of the materiality of a cathedral challenges the understanding of the facade. While on the outside ancient forms symbolised tradition and tell of nationalism as a political ideology, the reinforced concrete of the structure and the multi-ethnic construction workforce tell stories of modern construction technology and of traditions of co-operation. However, the motive to introduce such narratives of materiality cannot be a solemn rhetoric of revealing hidden truths. Instead the articles in this volume show how contested meanings and defiant practices on the construction site offer a focal point to write good and insightful histories. In this volume Patrick Leitner notes about New York's reconstruction that an inclusive world building process originated out of "contradictory elements of dignified harmony and material struggles". It was precisely the unplanned

15 Miller, Daniel, "Materiality. An Introduction", in: Ibid (ed.), *Materiality*, Durham et. al 2005, 1-50, 19.

16 From the poem "Fragen eines lesenden Arbeiters" translated by Michael Hamburger: Brecht, Bertolt, *Poems 1913-1956*, New York/London 1976.

17 See e.g. Ginzburg, Carlo, *The Cheese and the Worms. The Cosmos of a Sixteenth Century Miller*, Baltimore 1980.

and disjointed chaos in every corner of the city that created "the idea of a common whole."

When thinking about the past and the present of construction sites as a reservoir to understand the simultaneous building of worlds with heads and hands, the fourth topic we debated repeatedly was the characteristic of these places of production of being open and permeable as well as ephemeral. This spatial and temporal characteristic stresses again how valuable construction sites are for cultural studies, as their precarious status illustrates that the world we live in comes into being by discourses and practices that constitute a source of constant performative reconstruction. Christoph's article urges anyone who thinks and writes about construction sites to closely observe the stories one wants to tell and to be very specific about where your construction site is located in space and time.

In the lifetime of a street, a construction site normally only marks a passing moment. While houses are measured in decades, centuries and sometimes even millennia, construction sites seem to come from another order of temporality. They appear, change constantly and are gone after a few years to leave a finished building behind. Joy Parr is one of those who have shown that ephemeral changes can become an essential part of the study of major engineering projects. In a paper about a dam project in Canada she calls this slightly solemn *"the whispering of ghosts"*[18]. For the historian the question of sources becomes a pressing problem. When having to rely on printed, archival, personal and other documents, one can only keep an eye open and hope to find such elusive and well-hidden creatures.

While the construction of a house or even a city block rarely takes more than two or three years, there are exceptions. Germany's public is in a mixture of dismay, mockery and a severe case of national embarrassment when looking at the construction sites of an opera house in Hamburg and an airport in Berlin, which drag on for decades with exploding costs. The critics of Stuttgart 21 already see the infrastructure project as part of this shameful list. On one hand, the broad coverage tells of technical details on site that are otherwise invisible. On the other the situation seems to invite stories that follow the interpretations of hubris and nemesis. Monumentality, as Lynn Meskell diagnoses in a material culture study on the impressive remnants of Egyptian Culture, is the apprehension of the divine, but is also about the control of nature and society.[19] To look at construction sites of such monuments can tells of success, failure and the cost of such aspirations.

Most construction sites are less spectacular than the building of the Temple of Thebes. When I look out of my window today, I see two cranes behind the roof of the house opposite the road, I see their slow circle while sitting on the couch and sometimes I catch an operator climbing up or down to his lofty workplace. During the night the bright lit logos of the operating companies shine into the room. Through the open window the constant humming of machinery is pervasive. When I pass the construction site on

18 Parr, Joy, "The Senses and the History of Technology", in: *Technikgeschichte*, vol. 82, 2015, 11-26, 19

19 Meskell, Lynn, "Objects in the Mirror appear closer than they are", in: Miller, Daniel (ed.), *Materiality*, Durham et. al 2005, 51-71.

my way to the university or the supermarket I pay attention. I need to. The lorries are shunting on the road and the cyclists needs to carefully navigate through the huddle. The vans are from all over Europe, the languages one hears are too. It's a place of fast change, driven partly by the globalised and booming car industry in Stuttgart. This construction site also is a place with open borders, which are crossed by workers, engineers and material from near and far, a place of social differentiation and Europeanisation.

On the coasts of Spain or Greece a lot of construction sites seem frozen in time and space. I was on Fuerteventura last winter, in the small town Corralejo on the far northern end of that rock and desert island in the Atlantic. Half finished, the ruins of land speculation and the 2008 economic crisis were crumbling monuments right at the beach of the tourist [fig. 2]. The borders of these sites were barred off with high fences and crowned with barbed wire. On other places on the island, the abandoned sites were open to nothing more than the sand moving in and out permanently on windy Fuerteventura [fig. 3]. Many people have suffered, many artists have depicted such half build ruins in Europe's south. To catch a glimpse of how stepping on such places can feel, one can for example read *The New Yorker's* report "The Hangover" from 2013 with beautiful pictures by Simon Norfolk. With all that abandoned houses and construction sites, Spain has become a "museum of doomed development", the journalist Nick Paumgartner writes. He suggests understanding what to him seems like the remains of a "fireless apocalypse" as a way to "perceive an economic crisis."[20]

Both types of construction sites have a fascination of their own: The moving place, where big machinery tears down old structures, where workers move busily around and where cranes tower high in the sky; and the abandoned place, where failed human activity is decaying in its exposure to nature. Maybe the fascination of success, of growing buildings, of busy

Fig. 2 & 3 Abandoned construction sites on a Spanish coast (Corralejo, Fuerteventura, December 2013).

20 *The New Yorker*, 25 February 2013, http://www.newyorker.com/magazine/2013/02/25/the-hangover-2 (accessed 22-6-15).

activity, of large machinery mocking the puny strength of our bodies, but also the fascination of failure and decay, when abandoned construction becomes a still life displaying *vanitas* and other allegorical connotations, maybe all these things can be explained with a large and changing reservoirs of the sublime.

While Karlsruhe, Stuttgart's rival and sister city, currently celebrates its 300th anniversary, the building of a subway turns the historical roads of the centre into a single construction site. The museum ZKM installed a number of artworks around the town. In one of them, the artist Leandro Elrich lets the model of a historical building by the city's architect Friedrich Weinbrenner (1766-1826) hang from the cable of a crane over a central market square on which construction work is going on [fig. 4]. It really is a striking view and popular with tourists who take pictures in front of a house floating over the city under construction.[21] The roots that hang from the base of this installation let the house appear to be ripped out of its "natural" neighbourhood. The installation describes the construction site as a symbol of both *vanitas* and change that mesmerize the tourists. Fascination originally means bewitching, and the bewitched gaze is one that can only be turned away with effort. Even if the sublime has such pervasive cultural connections with construction work, the material site must be accounted for, too. Material obstinacy and the construction of power need to be considered,

Fig. 4 Leandro Elrich's installation hanging on a crane above the construction site of a subway in Karlsruhe, July 2015.

21 For an overview of the art program "The city is the star. Art at the construction site", see: http://zkm.de/en/presskit/2015/the-city-is-the-star-art-at-the-construction-site (accessed 14-7-15).

as ecological crisis or the brutal and harsh realities of slave labour at the construction sites of the Football World Cup in Qatar teach us in our time.

It can be very important where the temporal and spatial borders of a construction site are and how they are structured and perceived. These borders can make the difference between an imagined Eden and an apocalypse of failure and misery. Edited volumes like this one can only be snapshots out of a vast sea of research. As always, there are many more things to explore. When it comes down to the intention of the conference and this volume to explore the potential of construction sites as spaces where the world is built with head and hand, I am very happy I learned so much and met so many inspiring scholars. I am very grateful to everyone who participated at the conference and contributed a chapter to this book. Thank you for all the questions answered and the many new ones raised.

I owe thanks to a lot of people and institutions who made the conference and this book possible with their generosity. First there are my colleagues Thomas Schuetz, Sonja Petersen, Christine Etteldorf and Giovanni Brullo as well as the head of the department for the history of the impact of technology Reinhold Bauer, who gave support in every possible way. Elke Uhl and the IZKT (Internationales Zentrum für Kultur- und Technikforschung) of the University Stuttgart made the whole project possible with their financial and logistical help. Nora Heinzelmann's patience and her help in typesetting this volume were invaluable. Andrea Beck, Sandra Blaauw and other helpful people from the staff of Stuttgart city library were great hosts for our conference. I would also like to thank the Verein der Freunde des Historischen Instituts der Universität Stuttgart for their help and the James-F.-Byrnes-Institut for making it possible to invite a speaker from the US.

Eike-Christian Heine, Stuttgart, July 2015

LIST OF FIGURES

Fig. 1 – 4: Photos taken by the author.

PART 1: SOCIALISM UNDER CONSTRUCTION

SOCIALIST CONSTRUCTION AND THE SOVIET PERIODICAL PRESS DURING THE FIRST FIVE YEAR PLAN (1928-1932)

Marie Collier

1. INTRODUCTION

Much scholarly attention has been paid to the extensive economic, social and cultural changes that occurred during the First Five Year Plan (hereafter Five Year Plan) in the Soviet Union. However, the visual portrayal of socialist construction on the Soviet construction site has not been considered in depth.[1] Throughout the course of the Five Year Plan, photo-illustrated periodicals created clear visual links between the construction of buildings and the construction of socialism. In these mass circulated publications construction was represented as not only a reflection, or documentation, of the social reality of Soviet labour and production, but also as a metaphor for the successful building of socialist ideology amongst the population. While photo illustrated periodicals recorded the rapid physical and ideological transformations throughout the country, they also propagandised progress, ignoring or glorifying the hardships that workers endured on the way to industrialisation and socialism.

This paper will focus on two Soviet periodicals, *Nashi Dostizheniia* ("Our Achievements"), and *SSSR na Stroike* ("USSR in Construction") to explore how the use of photographic techniques, photomontage and graphic design assembled images of socialist construction during the Five Year Plan. Periodicals were an important means for mass education in culture and ideology as well as political indoctrination and therefore, their contents provide important insight into the ways in which Soviet values were transmitted to the public.[2] In these two magazines, construction sites are presented as equally if not more important than completed projects in order to portray the activities of the Five Year Plan as a continual process of production and advancement. At the same time, the mass circulation of photographs of construction sites drew attention to the process of Soviet construction as a product and an end in and of itself. It also presented construction as a product for mass consumption.

As the central medium for disseminating photographs to a mass audience, magazines contributed significantly to the construction of a shared culture and belief in Soviet socialism. Beyond their role in shaping a collective imagination, magazines provide a glimpse into the changing cultural values of the Five Year Plan and the development of the main components of Socialist

---

1 On the social and cultural changes during the Five Year Plan see Fitzpatrick, Sheila (ed.), *Cultural Revolution in Russia, 1928-1931*, Bloomington/London 1978; Clark, Katerina, *The Soviet Novel. History as Ritual*, Chicago 1981; Hoffman, David, *Stalinist Values. The Cultural Norms of Socialist Values, 1917-1941*, Ithaca/London 2003.

2 Wolfe, Thomas C., *Governing Soviet Journalism. The Press and the Socialist Person after Stalin*, Bloomington 2005, 1-33.

Realism.[3] When the photographs of construction in *Nashi Dostizheniia* and *SSSR na Stroike* are considered within the context of the Soviet mass media system at the time, the construction site becomes an important site of not only material but also cultural production.

2. THE FIVE YEAR PLAN AND CULTURAL REVOLUTION: BUILDING PROLETARIAN CULTURE

The Soviet government launched the first of several Five Year Plans in 1928.[4] It was a program aimed at rapidly increasing the Soviet Union's industrial and agricultural production. The main elements were the forced collectivization of agriculture and the planned industrialisation of the largely rural and agrarian countryside, which was to be achieved with a fast paced schedule of construction and production. It represented the consolidation of Stalin's power as leader of the Communist Party, and brought a reversal of the limited forms of private ownership and capitalism that had been established under Lenin's "New Economic Policy" (1921-1927). Collectivization and industrialisation were undertaken with the intention of establishing a socialist economic system that would result in fiscal and industrial self-sufficiency thus fulfilling Stalin's doctrine of "socialism in one country". Accordingly, this period from 1928 to 1932 is marked by massive social and physical changes across the landscape of the Soviet Union.[5] These included a migration of 23 million peasants from rural to urban areas between 1926 and 1939. Most joined the industrialised workforce, doubling its numbers from before the beginning of the Five Year Plan.[6] This migration created an urgent need to build the necessary infrastructure and housing for workers in addition to the ambitious scale of industrial construction. The rapid pace of change that resulted from migration, construction and collectivization led to housing shortages, famines and a dirth of skilled and educated labourers in many towns and cities.

The Five Year Plan fostered a cultural revolution that was aimed at replacing the pre-revolutionary bourgeois intelligentsia with a new elite formed of members of the working class. Party leaders hoped that industrialised labour would promote the "socialist transformation of the peasant"[7]. They believed that this could be achieved by training and educating members of the working class to lead and manage production.[8]

Along with strengthening the ties between the working class and Soviet intellectuals the Cultural Revolution ushered in a vigorous, often mil-

3 Socialist Realism was decreed the official form of Soviet artistic production in August 1934 at the Congress of Soviet Writers. For a history on its developments see: Gutkin, Irina, *The Cultural Origins of the Socialist Realist Aesthetic 1890-1934*, Evanston 1999.

4 The First Five Year Plan was announced in 1928 and officially began in 1929.

5 The Five Year Plan was completed in 1932, one year ahead of schedule and was hailed as a great success for achieving the targets so quickly despite the fact that many of the figures used to determine this were false and targets had not been met.

6 Hoffmann, David L., *Peasant Metropolis. Social Identities in Moscow, 1929-1941*, Ithaca/London, 1994, 2.

7 *Torgovo-promishlennaia gazeta*, 9 November, 1926 quoted in: Rassweiler, Anne D., *The Generation of Power: The History of Dneprostroi*, Oxford: Oxford University Press, 1988, 55.

8 Fitzpatrick, *Cultural Revolution* (fn. 1), 8.

itant, call for the development of a proletarian culture that both stemmed from and represented the worker.[9] Since the post-Revolutionary period avant-garde groups such as the Constructivists and Productivists had promoted the integration of art and life through the mechanisation and industrialisation of the creative process, and by urging artists to work alongside the proletariat.[10] The cultural revolution intensified this drive to bring art closer to the practices of everyday life. Cultural producers were urged to go as "brigades" to the many construction sites emerging throughout the country. The intent was for artists to embed themselves amongst the worker and record the progress being made. Workers were also encouraged to take an active role in the production of proletarian culture by learning skills in creative mediums such as writing, and photography.[11]

The quest for proletarian culture created an atmosphere of competition amongst different groups who jockeyed to position themselves as the true representatives of the workers. Organizations such as the "Revolutionary Association of Proletarian Writers" (RAPP), the "Russian Society of Proletarian Photographers" (ROPF), and *Oktiabr* ("October") adopted and followed aggressive cultural slogans for socialist construction. Cultural features and social roles associated with the broadly defined bourgeois intelligentsia were rejected.[12] In theory this meant the rejection of professional specialisation, and the replacement of experts with amateurs.[13] Therefore, traditional specialist artistic disciplines such as easel painting, and classical sculpture were dismissed as elitist and unrepresentative of proletarian interests. Similarly, by 1929 the non-objective geometric art of the early Russian avant-garde was increasingly condemned as "Western bourgeois formalism" and incomprehensible to the masses. In their place, forms of cultural production such as documentary photo journalism and concepts such as the "literature of the fact"[14] that purported to produce objective representations of reality emerged as proletarian alternatives.[15]

9 Taylor, Brandon, *Art and Literature Under the Bolsheviks: Volume Two, Authority and Revolution 1924-1932*, London 1992, 100.

10 For more on Constructivism and Productivism see, Lodder, Christina, *Russian Constructivism*, New Haven 1983; Gough, Maria, *The Artist as Producer*, Berkeley/London 2005.

11 Clark, Katerina, "Little Heroes and Big Deeds: Literature Responds to the First Five Year Plan", in: Fitzpatrick, *Cultural Revolution* (fn.1), 189-206; Tretiakov, Sergei, "*Ot fotoserii—k dlitel'nomu fotonabliudeniiu*" ("From the Photo-Series to Extended Photo-Observation"), in: Trans., Fore, Devon, *October*, 2006, 71-77. Originally published in: *Proletarskoe foto*, 4, 1931, 20–43.

12 Sheila Fitzpatrick and Moshe Lewin describe how the intelligentsia came to be broadly defined as one group that comprised a diverse set of individuals including intellectuals, artists, specially trained professionals such as engineers, and the pre-Revolutionary political and social elite.
Fitzpatrick, Sheila, *The Cultural Front: Power and Culture in Revolutionary Russia*, Ithaca/London 1992, 4-5; Lewin, Moshe "Society, State, and Ideology during the First Five Year Plan"; in: Fitzpatrick, *Cultural Revolution* (fn. 1), 57-73.

13 Taylor, *Art and Literature Under the Bolsheviks* (fn. 9), 100.

14 Writers Sergei Tretiakov, Osip Brik and Nikolai Chuzak proposed the *literatura fakta* ("literature of the fact") as an alternative to the psychological novel that was popular at the time. Instead of the fictional literary novel, newspapers, journals, travelogues and reports were to be the new form of writing that would better connect with readers, aid in the immediate construction of the new political and cultural consciousness, and better reflect the era in which they were living. Barooshian, Vahan D., "Russian Futurism in the Late 1920s: Literature of Fact," in: *The Slavic and East European Journal*, vol. 15, 1971, 38-46; see also, Chuzak, Nikolai (ed.), *Literatura fakta*, Moscow 1929.

15 It is important to note that realist painting also came back to the fore at this time with groups such as

The "proletarianisation" of culture was also tied to the notion of the formation, or the socialist construction (*sotsialisticheskoe stroitel'stvo*), of a New Soviet Man. The New Man would be reforged through the process of industrial labour, and would emerge as an ideal socialist worker and citizen. Consequently this period is marked by an enthusiastic drive to industrialise all aspects of Soviet life, including cultural production.

The popular language of the time reflects this attitude in the frequent use of the vocabulary of industrialism in relation to everyday life. Slogans such "Join the shock work on the cultural front!" convey the desire to industrialise the creative process and production, while the use of the words building, constructing (*stroitel'stvo, sooruzhenie*), and reforging (*perekovka*) in relation to human development indicate that people were considered parts of the machines that would drive industrialisation.

As a way to demonstrate their commitment to proletarian culture, creative producers were also required to take a class stance in all of their creative activities.[16] In visual culture, this was characterised by a requirement for artists to actively participate in the visual representation of socialist construction, or to create works that clearly contributed to the ideological education of their audience.

The term socialist construction had been in circulation from the early Revolutionary period in Russia as a phrase to describe the establishment of socialist institutions. However, from the late 1920s its ubiquity in mass media and its application to all aspects of Soviet life suggests an active desire to rebuild the whole of society along socialist lines.[17] Jeffrey Brooks has noted that from the beginning of the Five Year Plan, articles on socialist construction made up a quarter of the articles in the popular press and a fifth of all editorials. By 1928 it referred to not only the creation of a socialist society through a class war against the bourgeoisie, cultural education and the forging of new citizens through labour, but also to the physical construction of architectural and engineering projects that would foster and reflect a socialist lifestyle in their form and function. This link was made clear in the large number of publications in this period that featured images of construction sites, buildings under scaffolding and vast expanses of land being tamed by cranes. The construction of buildings that facilitated Soviet production, such as factories and industrial plants, were given particular attention in the popular press.

An emphasis on architectural development prevailed during this period as the Five Year Plan also marks the beginning of the Soviet Union's the first post-Revolution building boom. After years of primarily paper architectural competitions, and proposals, projects such as Magnitogorsk, the Dnepr Dam and AMO automobile plant, were initiated and mostly completed.[18] New housing, workers clubs, schools, cantines, mass bakeries and cultural venues also materialised across Soviet cities and villages. The architecture of the Five Year Plan developed out of period of experimentation

the Association of Artists of Revolutionary Russia, see, Taylor: *Art and Literature* (fn. 9).

16 Taylor, *Art and Literature* (fn. 9), 96.

17 Brooks, Jeffrey, *Thank you Comrade Stalin! Soviet Public Culture from Revolution to Cold War*, Princeton 2000, 38.

18 Construction on parts of Magnitogorsk and the Dnepr Dam continued into the late 1930s.

with new architectural forms and typologies in which architects and critics sought to define a specifically Soviet architectural language while also constructing buildings that would accommodate a collective lifestyle and contribute to the development of proletarian culture.[19] The circulation of architectural photographs in periodicals targeted at all segments of the population, from children to elite members of the Party, highlighted the importance of architecture and the construction of new buildings as an essential component of the foundation of socialism.

3. THE SOVIET PERIODICAL PRESS AND PHOTOGRAPHY DURING THE FIVE YEAR PLAN

During the Five Year Plan photography came to be recognized by the Communist Party as an ideological medium and an important tool with which to influence the collective imagination and to shape the masses' perception of reality. Over time, this function of photographs, and printed images in general, became an essential part of Soviet life.[20] The Soviet periodical press faced a dramatic rise in production after 1928 and photographic spreads were increasingly incorporated into publications throughout the early 1930s.[21] Thousands of new periodicals emerged which sought to raise literacy rates and the cultural level of the workers through simply designed visual layouts that could be easily comprehended by a broad range of viewers.[22] Though the role of posters, newspapers and novels in achieving this has been studied in depth, journals still remain largely overlooked in art historical studies. Nonetheless, they provide important insight into how socialist construction was visually communicated and circulated to a mass audience. Unlike a poster which would be shared publically in workers' clubs, outside factories, in city squares and civic buildings for people to view in passing, the illustrated journal allowed for private contemplative consumption of the images, texts and captions, and were therefore an important format for mass political education and the shaping of ideology. Furthermore, due in part to the high cost of securing copyright permissions to reproduce paintings, periodicals relied increasingly on photographs which were bought as ready made press-clichés. Different publications often purchased the same photographs and reprinted them in several different issues. They also frequently reprinted photographic spreads and individual photographs from special edition periodicals, posters, and albums in the more mass market publications.[23] This contributed to the formation of a unified visual language, and

---

19 On developments in Soviet Architecture see, Magomedov, Selim Khan, *Pioneers of Soviet Architecture. The search for new solutions in the 1920s and 1930s*, New York 1987.

20 Groys, Boris, *Art Power*, Cambridge (Mass.)/London 2008, 145-47.

21 In 1925 the USSR published 589 periodicals, by 1929 there were 1700 in circulation. Ovspian, P., *Istoriia noveishei otechestvennoi zhurnalistiki* ("The History of Modern Domestic Journalism"), Moscow 1996, 43.

22 A Central Party Committee resolution on 4 April 1921 declared that periodical press designers had to produce layouts that would be easily comprehended by the reader, "Tsyrkul'ar Tsentral'nogo Komiteta Parti ot 4 Aprelia 1921", in: Romanenko, Katerina, "Photomontage for the Masses. The Soviet Periodical Press of the 1930s", in: *Design Issues*, vol. 26, 2010, 29-40, 5.

23 Romanenko,"The Visual Language of Soviet Illustrated Magazines" (fn 22), 33-34.

an iconographic programme of socialist construction through architectural construction as certain images became very well recognised.[24]

As a result of the increase in photo-illustrated publications, Soviet citizens were inundated with images of socialist construction throughout this period. The reproduction of millions of photographs of construction sites shaped not only the political unconscious, but also the realm of the imaginary. Photographs were presented and described in the media as visual facts that accurately reproduced images of reality. Over time this contributed to the general understanding amongst the population that photographs were a reflection of reality rather than a framed representation of it.[25]

The periodical press also contributed to the development and dissemination of socialist realism as the only official format of Soviet creative production. As Elizabeth Papazian has summarised, socialist realism was based on a set of implicitly and explicitly defined aesthetic and rhetorical conventions to influence cultural behavior and build a coherent and comprehensive Soviet culture. Artists were required to present Soviet reality as a process that was advancing toward a historically inevitable establishment of socialist utopia. The numerous construction sites emerging across the country provided photographers and the periodical press with a good visual metaphor for this process of advancement. Through an emphasis on photographs as facts, the use of the language of documentary objectivity and uninhibited positivity, socialist realism showed Soviet citizens how to interpret the world around them.[26] Though it was not officially imposed until 1934, many of the conventions of image construction that would come to characterize socialist realism from the mid 1930s onward were developed in periodicals such as *Nashi Dostizheniia* and *SSSR na Stroike* during the Five Year Plan.

Professor of Soviet literature and culture, Evgeny Dobrenko has identified the reproduction of images of socialism as a key factor in the production of Soviet reality. He argues that the main purpose of socialist realism, with its emphasis on portraying reality in an overtly positive and optimistic manner, was to produce socialism itself. Dobrenko writes that the fetishization of production in Soviet mass media is equivalent to the capitalist fetishization of consumption and that this led to an emphasis in the media on the process of production, over the products of production. In other words, the product of Soviet production was socialist production (as opposed to exploitative capitalist production), which would result in the creation of a new man.[27] If we consider construction as a form of Soviet production, the same can be said for the emphasis on the construction of buildings over the completion of buildings as expressed in the periodical press during this period. In addition, within the drive to industrialise all aspects of Soviet life, the reproduction of Soviet culture and images of social, economic and political progress in periodicals was a form of production.

24 Ibid., 102.

25 Dobrenko, Evgeny, *Political Economy of Socialist Realism*, New Haven 2007, 6.

26 For the relation between the use of documentary and socialist realism see, Papazian, Elizabeth Astrid, *Manufacturing Truth. The Documentary Moment in Early Soviet Culture*, Dekalb 2009, 9.

27 Dobrenko, *Political Economy* (fn. 25), 256.

Dobrenko takes his analogy further by arguing that in the way that capitalist production creates a consumer, the production of images in the Soviet Union created a reality.[28] He claims that "in a situation wherein reproduction is production, the more of it that is represented, the more of it is produced"[29]. That is to say, the creation and reproduction of images of construction was equally important as the physical construction that was underway. It was a form of Soviet production that was contributing to socialist construction. Furthermore, while purporting to reproduce documentary reflections of the social reality of Soviet construction, periodicals were in fact producing their own reality which corresponded with the desired image of Soviet progress as prescribed by the Party.

Thus through mass circulated photographs in periodicals such as *Nashi Dostizheniia* and *SSSR na Stroike*, socialist construction became a pictorial as well as a literal concept that captured the changing social, political and visual landscape and conveyed it back to the workers. Construction sites were represented as places where workers overcame many challenges – in the landscape, their education, political awareness, economic stability and social security – to transform themselves, all the while building the physical structures that would ensure the continued improvement of their society and the advancement of socialism. They were also presented as places of constant change and progress for both the workers and the landscape. The frequency with which photographs of construction sites and their thriving workers were represented in combination with similar themes in literature, theatre and film, functioned to create the collective imagination of socialist construction. This helped to produce optimism and faith in the success of the Five Year Plan, and socialism in general, during a very unstable time.

In their representation of socialist construction, periodicals relied heavily on the documentary capabilities of photography in order to establish a visual language that would appeal to a mass audience and portray convincing realism. Photography was the medium most favoured by the Party and magazine editors for this purpose, for its perceived ability to reproduce an objective image of reality efficiently and effectively. Its mechanical properties also resonated with the spirit of industrialisation at the time. The critic Grigori Boltanski summarised the dominant position on photography as a persuasive and practical medium writing in the magazine *Sovetskoe foto* ("Soviet photo") in 1926 that photo-reportage or documentary photography in particular was considered "socially important", and was regarded as, "the most significant social need demanded by the victorious proletariat"[30]. Boltanski later declared in *Pravda* in 1931: "The clearness and convincing concreteness of the photograph in unity with the accompanying caption give to the hands of the press an additional weapon for the mobilisation and organisation of the masses." He argued that the advantage of photography lay in its uniquely "concrete" and "irrefutable language" that was "accessible

28 Ibid., 7.

29 Ibid., 263.

30 Boltianski, Grigorii, "Fotograficheskaia zhizn' v sovetskii" ("Soviet Photographic Life"), in: *Sovetskoe foto*, 6, 1926. Quoted in: Elliot, David, *Photography in Russia, 1840-1940*, London 1992, 59.

to the broad millions of masses"[31]. Several critics and photographers echoed these sentiments throughout the Five Year Plan and the belief in photographs as powerful tools for socialist construction was widespread amongst many different, even opposing, groups of creative practitioners. Additionally, photography was considered to be the only medium capable of keeping up with the rapid pace of technological and social change occurring at the time.

The format of the magazine was particularly well suited to the task of representing construction as a process, as it allowed a visual narrative of progress to unfold page after page, or even volume after volume. Magazines combined photographs with text and captions to shape the desired interpretation of the images. Though illiteracy rates had dropped significantly since the early 1920s, a focus on the visual presentation of Soviet ideals was still prioritised as a means to educate workers and peasants.[32] The persuasive capabilities of the photograph were further expanded with the introduction of the photographic essay (*fotocherk*) to the Soviet periodical press in 1923. Photographic essays could send ideologically "correct" messages in a subtle manner, through the use of serialised photographs to create a narrative. The serialised images, combined with text, avoided blunt propaganda by exploiting the documentary quality of photography. They embedded the master narrative of progress and social advancement in stories of specific construction projects, people, or places that were captured by the camera, often in diverse places at different times, and assembled in the magazines. The realism conveyed by the photographs in combination with the text, made it difficult for the reader to discern their contrived nature.[33]

Periodicals regularly employed graphic design techniques, such as asymmetrical layouts, the juxtaposition of images and text, and the use of different scales. They also frequently employed photomontage, which encorporated all of the techniques above, to create engaging and ideologically convincing spreads in magazines. The technique of assembling different photographs together on one page had been developed in Russia and Western Europe in the early 1920s and became a particularly persuasive tool in Soviet printed media.[34] Throughout the 1920s and into the early 1930s the term photomontage described photographs arranged in a sequence, one below the other on one page, similar to a photographic essay, as well as the technique of cutting up and re-arranging photographs into a new composition. Both techniques were thought to be able to convey the appropriately socialist "dialectic materialist" approach to image construction.[35] While an individual photograph on its own presented too

31 Boltianskii, Grigorii, *Ocherki po istorii fotografii v SSSR* ("Essays on the History of Photography in the USSR"), Moscow 1939, 98.

32 Brooks, *Thank you Comrade Stalin!* (fn. 17), 56.

33 Romanenko, *The Visual Language of Soviet Illustrated Magazines* (fn. 22), 143.

34 Though it falls beyond the scope of this paper to address the artistic debates that surrounded the use of photomontage in periodicals during this time, it is necessary to point out that it was considered a useful, if contentious technique.

35 The terms *fotomontazh, fotolitho montazh, and fotocherk* were used interchangeably to refer to both serialized photographs and photomontages until the mid 1930s. Romanenko, *The Visual Language of Soviet Illustrated Magazines* (fn. 22), 139-140; On photomontage and the dialectic materialist creative method see, Gassner, Hubertus, "Heartfield's Moscow Apprenticeship, 1931-1932", in: Pachnicke, Peter/

narrow and specific a view, a photomontage could provide the necessary socialist context, by incorporating both the general and the specific components of a given subject.[36] This was achieved by either combining images from diverse sources together on one page, or assembling a series of whole images that contained both broad shots and close-up details.

As the Five Year Plan progressed and Soviet visual culture moved closer to socialist realism, cultural producers were increasingly required to achieve a fine balance of conveying both documentary specificity and general typicality in everything they created. This is notable when surveying both *Nashi Dostizheniia* and *SSSR na Stroike* as their use of photomontage increases toward 1932, despite a rise in attacks on the formalist nature of the technique that were waged in the press from 1930 onward. The continued use of the technique demonstrates that its visual qualities were still considered to be effective tools for constructing images as long as they served a clear ideologically didactic function. Arkadii Shaikhet and Simion Fridliand, photographers and members of the group *ROPF*, who led the attacks on "western bourgeois formalist montages" of rival group *Oktiabr* summed up the dominant position on the use of photomontage in 1931, writing that they allow "the presentation of facts in [a form that would be] supremely simple, easily accessible, and, at the same time, supremely impressive to any viewer"[37]. Editors continued to use graphic layouts and photomontage throughout their photographic essays to ensure the correct interpretation of the images. The juxtaposition of images, use of scale and text guided the readers' interpretation and reinforced the communication of the socialist realist programme of unquestionable optimism and realism.[38]

Furthermore, the obviously constructed images that result from the use of photomontage, reiterate the idea of a society that was under construction and being pieced together through labour and technological and industrial advancement. The combination of photographs in a photomontage could collapse time and space to illustrate the grand process of change taking place throughout the country over a long period of time, thus making it an ideal visual technique for representing socialist construction. Photomontages could combine representations of a work in progress with a finished project to convey the successful process of advancement that was being achieved through socialist labour with the need to continue working in order to accomplish the ambitious production targets that had been set.

4. *SSSR NA STROIKE* AND *NASHI DOSTIZHENIIA*

SSSR na Stroike and *Nashi Dostizheniia* represent two different extremes of the Soviet publishing industry. In writing about these two periodicals together, the aim is to demonstrate that socialist construction through photographs

Honnef, Klaus (ed.), *John Heartfield*, New York 1992, 269.

36 Gassner, "Heartfield's Moscow Apprenticeship" (fn. 35), 269-70.

37 Fridliand, Simon/Shaikhet, Arkadii, *Fotovystavka zhurnala 'Ogonek'* ("Photo-displays of the journal 'Ogonek'"), Moscow 1930, 3-4.

38 Romanenko, *The Visual Language of Soviet Illustrated Magazines* (fn. 22), 181.

of construction sites permeated all levels of society and pervaded all types of publications. Both were monthly magazines created at the beginning of the Five Year Plan and published by the state publishing house OGIZ-IZOGIZ. They were conceived and edited by Maxim Gorki following the success of the Soviet Pavilion at the 1928 Pressa Cologne Exhibition.[39] This was an international showcase of modern press, advertising and publishing at which the Soviets sought to display their progressive society with an elaborate exhibition designed by El Lissitzky. The Soviet display at the fair was commended by the international press for its innovative use of photography, photomontage and graphic design.[40] Both magazines shared the same editorial concept: to illustrate the achievements of the Five Year Plan to a mass audience by reproducing the "facts" of socialist progress, and they shared roughly the same circulation numbers; between 50-100 000 copies depending on the issue.[41] However they were created in different formats for vastly different audiences.

SSSR na Stroike was originally produced as an illustrated supplement to *Nashi Dostizheniia* in 1930 and quickly developed into its own publication. It was a lavishly produced, high quality, large-format magazine that was published in Russian, English, French and German. It was intended for the emerging managerial class and for foreigners and placed an emphasis on photographs over text. Early editions featured very little text at all.

Nashi Dostizheniia on the other hand, was a smaller, literary magazine, aimed at workers. It featured *ocherki*, essays that combined documentary facts such as statistics with literary narrative. As Gorki stated in his opening editorial, the magazine was "About the Little People and their Great Work"[42]. It was created exclusively for Soviet readers, and especially those who Gorki believed still needed to be convinced of and made aware of the many great achievements being made daily at the hands of the workers.[43] It featured fewer illustrations than *SSSR na Stroike* but often reprinted series of photographs and photomontages that were found in the more luxurious magazine or other state publications. However, the material in the magazine's archives in Moscow reveals that it had its own in-house photographers and photomontage artists and that publishing original photographic material was an important part of its editorial concept.[44] Despite their differences, both magazines openly expressed their mandate to promote the positive aspects of the Five Year Plan, emphasizing the success of socialist construction, and leaving out any negativity or self-criticism.

39 For more on the origins of the journals see, Wolf, Erika, *USSR in Construction. From avant-garde to socialist realist practice*, PhD Diss.: University of Michigan, 2008.

40 Aynsley, Jeremy, "Pressa, Cologne, 1928: Exhibitions and Publication Design in the Weimar Period", in: *Public Photographic Spaces: Exhibitions of Propaganda, from Pressa to The Family of Man 1928-55*, Barcelona 2009, 83.

41 Circulation numbers drawn from the back covers of the periodicals. The circulation figures for the foreign language editions were significantly lower, typically between 6-10000 issues.
Circulation numbers do not indicate readership as many copies would be sent to libraries, workers' clubs and factories where one copy would be read by several different people.

42 Gorki, Maxim, "O malen'khikh liudiakh I velikoi ikh rabote", ("On the Little People and Their Great Work") in: *Nashi Dostizheniia*, vol. 1, 1930, 1-5.

43 Ibid.

44 *Rossiyskii Gosudarstvennyii Arkhiv Literaturii i Iskusstva* ("The Russian State Archive of Literature and Art", RGALI) f. 617, op.1 ed. X 246, 85-90.

The process of construction was followed closely in both periodicals for several prestigious projects, particularly the Dnepr Dam, which was reproduced thousands of times in periodicals, posters, children's books, photographic albums, newspaper articles and postage stamps. *Dneprostroi*, as the construction project was called, was the largest hydroelectric dam in the world at the time of its completion. Once it was opened, it would generate power from the Dnepr River through a system of locks and a large power station that sat on top of protruding concrete sluice gates. Tens of thousands of labourers flocked to the site in the Ukraine in search of work and an opportunity to gain training and take part in the colossal construction project. The dam consisted of forty-seven sluice gates mounted between piers, nine turbines and generators that produced 16.7 million kilowatts of electricity, enough to power the numerous industrial plants that were constructed on site as well as the new city of Zaporozhe and the surrounding countryside.[45]

SSSR na Stroike featured the project twelve times over the course of four years. In the first example published in May 1930, the construction site is depicted as an expansive space that has been overtaken by machinery [fig. 1]. Photographs show the extent of the construction across the landscape, the chaos of building and the scale of the work being done that included building rail lines, pouring concrete for the sluice gates of the dam and the construction of new housing for the workers. As the first Soviet construction project to involve the widespread use of motorised cranes, which had been recently imported from the United States, there is little wonder that the media focused its attention on such an immense undertaking. It provided an excellent example of the benefits of the mechanisation of the workforce by illustrating the rapid pace and grand scope of the construction. Throughout this issue

Fig. 1 USSR in Construction (*SSSR na Stroike*), no. 4 1930.

45 A steel factory was in operation by 1933 and a coke factory, iron works and aluminium factory followed suit. Rassweiler, *The Generation of Power* (fn. 7), 11.

the machines are prioritised over the workers, reflecting the industrialisation drive and the integration of machines and everyday life. As Erika Wolf has argued, the cranes appear to operate independently of the worker and recall the prevailing slogan of the late 1920s, "technology determines everything"[46].

The sequence of images through the magazine along with the captions conveys a clear message of progress at the hands of industrialised labour. As readers turn each page they are confronted with new developments in the construction of the dam and surrounding facilities for the workers. The construction site is depicted in various stages of development from breaking ground to the completion of the first essential components of the project. Notably, these are the buildings that would house the thousands of workers who had recently arrived on the site from across the Soviet Union.

Throughout the issue photographs link the transformations of the landscape with the transformation of the daily lives of workers by incorporating images of workers being actively reconstructed along socialist lines through education, leisure, social, and cultural activities inside the completed buildings designed for socialist construction [fig. 2]. These include a theatre, a mass dining hall and new housing units that are placed alongside a photograph of a family at home reading the newspaper [fig.3]. By placing these photographs together in a two-page spread, the link between the newly literate, cultured, worker and the new architecture is made clear. The progress of socialist construction and the improvement of the lives of workers are visually equated with the progress on the construction site.

This combination of photographs of the chaos of the construction site at the beginning of the issue, with the completed buildings towards

Fig. 2 USSR in Construction (*SSSR na Stroike*), no. 4 1930.

46 Wolf, *USSR in Construction* (fn. 39), 200.

Fig. 3 USSR in Construction (*SSSR na Stroike*), no. 4 1930.

the end, emphasises the process of socialist construction as productive and successful. The completed buildings indicate that progress is being made, that industrialisation is under way and that socialism (as well as architecture) is being built. However, they also suggest that construction is ongoing. Though many of the buildings are finished, the photographs reveal that they still sit within a construction site as evidenced by the surrounding landscape of mud and piles of bricks and other construction materials.

The ongoing nature of construction during the Five Year Plan is further reinforced in the October 1932 issue of the magazine. It was designed by El Lissitzky, and devoted to the official opening of the completed dam. However the pages of the magazine are still dominated by images of construction [fig.4]. Lissitzky adhered to the increasing demands by critics and Party officials that images convey both the general and specific elements of the subject by using photomontage extensively throughout the issue. Though the purpose of the issue is to celebrate the project's completion, the pages of the magazine continue to highlight the construction site. The construction is shown as a taming of the land and overcoming the struggle in "the offensive" to produce an impressive, functioning hydroelectric dam and an entirely new town in what was once considered to be the middle of nowhere.[47] The way in which this is achieved differs notably from the 1930 issue and reflects not only the changing cultural priorities, but also the status of the worker after years of successful socialist construction.

Rather than emphasise the mechanisation of labour through a focus on technology and anonymous labourers, this issue depicts workers as distinct individuals who have been socially transformed through their

47 No author, "Introduction", *SSSR na Stroike*, 10, 1932, 2.

work on site. The text highlights this by revealing personal anecdotes about workers who had arrived on site illiterate, poor and unhealthy and who have emerged as successful builders, and leaders in their community through their engagement with politics and social activities. The celebration of individual workers through the inclusion of their photographs in several montages reflects the cultural shift as well as the success of socialist construction and the Five Year Plan [fig. 5]. From the 1930s onward, the emphasis on the political and social transformation of individual workers on the construction site was a common feature in publications such as *SSSR na Stroike* and *Nashi Dostizheniia* and it appeared in several other issues of both journals.

Fig. 4 & 5 El Lissitzky, photomontage, *SSSR na Stroike*, no. 10 1932.

Toward the end of the Five Year Plan, Stalin initiated a call for a more human conception of labour when in 1931 he declared: "The reality of our programme is active people, we ourselves"[48]. This shifted the focus away from technology (as it was expressed in issue 4 1930), and placed it onto the worker who became the central agent in socialist construction. It also signalled that a new Soviet worker has emerged from the process of socialist construction. It is reflected in the photomontages in this issue in the large scale of the workers relative to the dam and their placement in the foreground of most compositions [fig. 6].

Throughout this issue, the construction site and completed buildings were consistently juxtaposed together in one montage. This reiterated the importance of construction as a productive force in society that was continuing, yet making progress, satisfying the socialist realist demand to depict Soviet reality as a process of advancement toward Utopia. Both stages of the construction process are effectively equated to one another by their placement on the same page. Furthermore, the photographs document the process and progress of the construction project over time, while presenting it as a finished product ready for instant consumption by the readers.

Nashi Dostizheniia's editors made use of photomontages for a similar purpose to both sum up the successes of the First Five Year Plan and introduce the Second.[49] A series of photomontages reprinted in *Nashi Dostizhniia* in October 1932 display the progress on a range of recent achievements in socialist construction and production. The entire issue, designed by Varvara

Fig. 6 El Lissitzky, photomontage, SSSR na Stroike, no. 10 1932.

48 Stalin, Joseph V., "'New Conditions: New Tasks in in Economic Construction.' A speech delivered at a conference of economic Executives on 23 June 1931", in: *Problems of Leninism*, Peking 1976, 559. Originally published in *Pravda*, 183, 5 July 1931.

49 The Second Five Year Plan was launched in 1933 with a focus on heavy industry.

Stepanova, the partner of the constructivist artist and photographer Aleksandr Rodchenko, combined photomontages with ocherki and strong graphic panels conveying statistics of the Plan's success. The eight photomontages in the issue reprinted from an album of photographs by Boris Ignatovich, Elizar Langman, and Georgii Shurikhin, depict themes of agriculture, collectivization, education, architecture, metallurgy, the construction of hydroelectric dams and the military. They are distributed at even intervals throughout the magazine. Though they are each captioned with quotes from Stalin, they do not correspond directly to specific articles in the magazine. Instead, they provide a visual accompaniment to support the theme of the issue.

Toward the beginning of the issue a full page photomontage places construction in the centre of the larger goals of the Five Year Plan [fig. 7]. A quote from Stalin, in the bottom left corner, from 1931, reads, "We are lagging behind the advanced countries by 50-100 years. We must make up this distance in ten years. Either we do it or are crushed"[50]. A photograph of a beaming Stalin staring out to the reader occupies the centre of the image. The construction behind him is a sign of the progress being made. The buildings under scaffolding recede into the distance, and create a visual divide between the central themes of industrial and agricultural production. In the foreground, newly manufactured Soviet tractors are maintained by eager kolkhozniks (collective farm workers), while combines plow the fields to the right. Collectivization is well under way and thriving, and it is thanks to the construction of the factories that produce tractors, and the industrial plants such as those shown in the background. Furthermore, the aerial view shows the construction of worker's housing progressing in the right hand corner.[51]

Fig. 7 *Nashi Dostizheniia*, no. 11-12 1932, photomontage by Boris Ignatovich, Elizar Langman and Georgii Shurikhn.

50 *Nashi Dostizheniia*, 11-12, 1932, 8.

51 The bird's eye view photograph of housing construction in the outskirts of Moscow is one that was frequently re-published in different formats for theatre backgrounds, other photomontages, and large street decorations throughout the 1930s. It was originally part of a photomontage by John Heartfield, designed for *SSSR na Stroike* and published in issue no. 4 1931 devoted to the reconstruction of Moscow.

This montage is followed by another that replaces Stalin as the central figure with an image of workers as the agents of socialist construction, reflecting once again the shift in attitudes toward the agency of the workers, while simultaneously linking Stalin with the workers as they each occupy the same central position in different compositions, in this issue.[52] Newly completed projects are positioned behind the workers, while the diagonal cut of the page creates a juxtaposition of before and after. The construction sites in the foreground allude to future work to be done, emphasising the beginning of a new cycle of construction for the Second Five Year Plan and its focus on heavy industry.

Toward the end of the issue, urban architecture is highlighted with a two-page spread that depicts a variety of workers' housing [fig. 8]. All of the housing units are under construction. In the top to left hand corner, there is a close-up photograph of a brick-layer in action, actively contributing to socialist construction through the labour of architectural construction. The caption reinforces this reading of the image stating: "The daily care of the material needs of the workers and peasants is the most important condition for building socialism"[53]. Here the worker is fulfilling the conditions for socialist construction, by actively engaging in physical construction of a building.

Fig. 8 *Nashi Dostizheniia*, no. 11-12 1932, photomontage by Boris Ignatovich, Elizar Langman and Georgii Shurikhn.

52 The caption reads, "Metallurgy is the basis of the economy. Furnace-men distributers, industry workers! Bolsheviks rapidly achieve a victory in the great struggle for metal!"

53 Author unknown, *Nashi Dostizheniia*, 36.

5. CONCLUSION

The reproduction of visual representations of the construction site played an essential role in the production of Soviet reality, in both the material and imaginary realm. The consistent presence of images of construction sites in Soviet periodicals throughout the Five Year Plan signalled the importance placed on the 'things to come', and the extraordinary planning taking place within the Party to ensure the successful completion of the Plan. Through the production and reproduction of these images, Soviet periodicals succeeded in circulating an idea of construction and socialist labour as a product for mass consumption. Magazines therefore functioned as both tools of persuasion and reflections of the culture at the time. By fixing the process of construction in photographs and photomontages, images became a product for mass consumption and one that contributed to the definition and visual expression of socialist construction.

LIST OF FIGURES

Fig. 1 – 8: Private Collection.

BUILDING LAND: CONSTRUCTED FOR OBLIVION

THE REPRESENTATIONS OF BUILDING SITES IN GDR FILM HISTORY

Marius Böttcher

1. THE PLACE WHERE ALL THINGS COME TOGETHER

Die Architekten ("The Architects", Kahane, 1990) was not only the last film produced in the GDR, it also was the least possbile. Written in the late eighties before the fall of communism and the GDR, the film offers a melancholic view of a time before events of history overtook everything in its wake. By the time it was released in the cinemas, change was well underway in the GDR and the film – after a long fight against censorship – was already deemed passé. The film setting is a construction site where construction never begins. The anticipated buildings are only ever visible in mock-ups or blueprints. It is through this motif that the film tells the story of the end of an utopia that was traditionally visualised by classic illustrations of "workers building socialism".

Die Architekten starts with a picture of fallow land and the enthusiastic hope of breathing life into this formerly barren landscape. The potential for this land appear endless. The miserable failed appearance of the construction site is itself a metaphor for progress denied, development stuck in a state of standstill, and the impossibility and impracticality of life and work in the GDR. *Die Architekten* can be seen as a story about the end of the idea of the GDR, which likewise started by breathing new life into a barren and destroyed country. However, one question remains: what happened?

In the history of East German cinema, construction sites were omnipresent settings for everyday stories. The country was imagined as a building land. There are several dozen films about reconstruction and progress, well known "proletariat" films or even post-industrial milieu studies demonstrating the interdependency of work and society. Some of them enjoyed big box office success; some were forbidden; and others were simply forgotten. Most of these scenes cannot be described as a motif, given that the construction sites were only shown as a common spectacle in the cityscape. However, it is not only the *explicit* representation of the construction site itself that makes these films interesting for cultural studies. Construction is one of the oldest cultural techniques and the construction site itself serves as a platform for the values of a society to come together in a formation of (sometimes) immense production, often starting on the clean sheet of an empty field. With these reflections in mind, the building site becomes an extraordinarily interesting place within the context of the GDR's history. From the moment of the GDR's inception, the construction site was not merely a place where things were built. The representation of building sites in the GDR offers various perspectives, each with multiple layers of meanings, and creates a specific "concept of self" of the workers' state. Volker Braun, one of the best known scholars of the GDR's contradictions and the country's aspirations for socialism, understands the operation of the construction site as a symbol machine of the GDR:

> People produce something big out of nothing as well as themselves in masses. At the same time they construct a new state which needs that kind of people. [...] However, our topic was not technology, but only humanity. So, the building land was a growing, a contradictory landscape full of human relation.[1]

In this understanding the "communist utopia of a de-estranged labour as well as the comprehensive renewal of the individual and the participation in a free socialist society"[2] find themselves in pictures. The construction site is explicitly rendered as a place where socialist ideas are constructed. Through symbolically powerful pictures of cities being rebuilt, these images show the success of collaboration and working together, and represent the working class and its influence positively. Mass production represented the assembly of collective work in practice and emphasized the non-individuality of workers. By depicting the power of the working class as *one* entity, and by connecting people into one successful and powerful unit, the images of the construction worker became a symbol for progress and an expected better future by collective labour. This philosophy was already established in several political strategies that incorporated the interactions of living conditions, labour and the flexibility of individual development. In these strategies, culture became an important tool, because it was seen as a useful synthetic identification pattern for the socialisation of the individual and its submission – or at least to lead the individual to its place in society and its role within the working class.[3] This, of course, already implies constructing, building-up and assembling. For this reason, the construction site in the films of the GDR became a protagonist itself that proclaimed the GDR's agenda.

This important aim of socialist art, influenced by the utopian idea of building up a *new* society, had its own history, predating the GDR. By September 1944, towards the end of the war, Wilhelm Pieck, the leader of the Communist Party of Germany, as well as Hans Rodenberg and Johannes R. Becher, two of the leading figures in culture and art, met in Moscow to debate the question of redefining the national self-concept of the German people. As Becher stated "total-defeat needs total-criticism in all fields"[4], he saw the urgent need for an intense discussion

1 Braun, Volker, "Interview mit Bernd Kolf (1974)", in: Ibid., *Texte in zeitlicher Folge*, Vol. 4, Halle-Leipzig 1990, 319 [this and all following translations to English by M. Böttcher].

2 Brosig, Maria, *Es ist ein Experiment. Traditionsbildung in der DDR-Literatur anhand von Brigitte Reimanns Roman Franziska Linkerhand*, Würzburg 2010, 185.

3 E.g. Arbeitsgruppe Kulturtheorie in der Sektion Ästhetik und Kunstwissenschaften der Humboldt-Universität zu Berlin, *Der Beitrag von Marx und Engels zur wissenschaftlichen Kulturauffassung der Arbeiterklasse*, Berlin 1970-75, 8. Online: http://www.kulturation.de/_bilder/pdfs/2012-03-26_Marxkultur.pdf (accessed 28-4-15).

4 Mückenberg, Christiane, "Zeit der Hoffnungen. 1946 bis 1949", in: Ralf Schenk (ed.), *Das zweite Leben der Filmstadt Babelsberg. DEFA-Spielfilme 1946-1992*, Berlin 1994, 8-49, 9.

about fascism. This led to the re-evaluation of German history, even prior to 1933.[5] The essence of this discussion was the strife to an anti-fascist democratic transformation, which followed the unresolved legacy of the failed 1848 revolution. The focus placed on 1848 was effectively an attempt at repressing everything that had recently happened – despite the fundamental and benign aspiration of overcoming the Third Reich. Such predefined understandings influenced and even determined artistic work so heavily, that it was fundamentally integrated into the infrastructure of film productions.

After the war, film production in Germany was controlled by the Allies in order to re-educate a population which formerly followed a fascist ideology. Instead of continuing film production by a variety of smaller companies like in the Western Zones, the Soviet occupation zone founded the *Deutsche Film AG* (DEFA), which aimed for a determined "re-start" of German film, especially in a specific aesthetic language. In the sense of re-education, these productions quite clearly had an ideological slant, rooted in the communist vision of fusing political practice and art. Artistic anticipation and political struggle fused together so that in the interpretation of the Marxism-Leninism, culture and society are thought as one. In this framework, art has the function of shaping the development of history and society, in order to vanquish the old ideologies (fascist and capitalist) and to build a new order.[6]

Thus, East German art was always proletarian, revolutionary and concept-driven in its own unique way, whereby these objectives, obviously, also became the foundations for film production.[7] But the aims of the DEFA-films were not only political or even propaganda-based. In many ways they were deeply rooted in a vision of something completely new. This concept of "a revolution in the image of building a new country and a new society" could first be seen on the urgently needed building sites of the reconstruction process in post-war Germany, remedying the large-scale destruction caused by the war.

2. ASSEMBLING THE MATERIAL

Early official documentaries shot by the DEFA were about post-war reconstruction. The first one, *Berlin im Aufbau* ("Berlin Under Construction"), was directed by one of the film institute's founders, Kurt Maetzig. His work

5 Ibid., 9.

6 Even their aim was dissociation by politics as well as by artists, there de facto still exists continuity especially in artistic expression. Kurt Maetzig wrote: "Despite all substantive opposite and intellectual dissociation, the appearance was similar to the former Ufa-films. We understood the dissociation in two directions: Firstly in the opposite of resilient- and military movies, which the Ufa produced during the war, and secondly the dissociation from pure entertainment films far removed from the real social life, the purpose of which was diversion. But for us, it seemed as we were initiating a radical break. In reality however, things were somewhat different. Many tradition lines, either negative or positive, were about this and continued for a long time." (Maetzig, Kurt ,"Neuer Zug auf alten Gleisen", http://www.filmportal.de/material/kurt-maetzig-ueber-die-ufa-tradition-und-die-defa-gruendung (accessed 28-4-15)).

7 Enshrined in the principles of so-called "Socialist Realism", the aesthetic purpose was to further the goals of socialism and communism accompanied by the aspiration of a true and faithful artistic representation of "concrete" history, in addition to re-building and re-educating the working class.

was followed by Richard Groschopp, the director of the 1946 production *Dresden*. Groschopp created a mixture of documentary material and fictional elements, juxtaposing images before and after the war in an effort to contrast them. Assembling and collating these materials into educational units and combining historical documentation (images of destruction) with the current reconstruction efforts created an intentional "construction" of history.

The film starts with a free camera move showing an abundance of figural decors and a seemingly unnecessary sculpture, embedded in a dreamworld of sandstone. Different perspectives of foreground and background, superimposed with layers of the past, produce a variety of historical epochs and overlapping worlds. All these images are old impressions of the baroque buildings of Dresden's old city, displaying the distant world of the eighteenth century. Moving shots taken from cars and the funicular railway connect the new times with the old and represent the continuity of history. However, this continuity is in fact synthetically fabricated. Just as the baroque buildings are a dreamland, once constructed to ban the *Daseinsangst*, an almost melancholic trial of that epoch to escape from the dark side of human existence, there is always a repression, a struggle to forget a time before everything lay in ruin.

After a hard break the camera stands still. It shows only the wreckage of the city and the dust clouds obstructing the view. Ruins collapse, and the dust, dirt and rubble create an opacity of material stratification [fig. 1]. Music no longer accompanies the scene; just empty, analogue noise. The viewer feels continuity and discontinuity, raw substance, and termination. Indistinguishable and mundane rubble clouds up the screen in a paradox of horrifying emptiness. Tilting layers of ruins, new vistas where formally

Fig. 1 *Dresden* (Screenshot, GDR, 1946).

buildings obstructed the view and constructed perspectives created by the emptiness, well planned views enhancing the impression of openness that establish a feeling of space, piled and accumulated history, stony remnants, loose odds and ends, indefinable city structures which do not indicate any inside or outside, no roofs or windows or any definable form, just lumpen material, all this and more.

The former imposing buildings of the baroque era are shown as crippled creatures, ragged and hollowed, as naked invalids, disabled from continuing their former baroque life. The speechlessness in the face of the overwhelming absence of things is felt and culminates in one camera tilt from the bottom of an empty base, once occupied by a statue, to the emptiness that has taken over its position. There is only the sky left, largely unchanged. But since Michel Serres, we know that the physical element of air cannot have a history because it is not capable of having any form of memory. The shot of the immutable sky is the symbolic and desperate attempt to forget the horror. However, if we take a look at the ground of the cities, all the material of the ruins have to be the fundamental starting blocks of the new. The material of the old city is the beginning of the new, so that there was no possibility of a genuine "re-start": it was about rebuilding, not building upwards or anew. In that sense, assembling means putting the old things together and taking from them some elements as new.[8] That is just as visible in the material nature of the scenes as well as the human resources utilized. From these films on, the building-up process and humanity were thought as one by the films of the DEFA.

Motifs of union and the masses show how the individual became part of an ensemble. The rhythmic movement of the workers, functioning as a single entity, synchronises their bodies in a rather mechanical way, bringing different materials together, just as the processes of compiling and merging materials at the construction site do. This imagined mode of assembling is the directing concept of the first period of construction site films of the GDR. Constructing socialism is portrayed as the building of a city and a community, forging material as well as human beings. The depiction of an absence of hierarchy and the creation of an image of the masses as a single subject, present in these films, display the political and social construction in a single place as the realisation of an utopian idea in an overtly symbolic, almost pathetic way: the masses, machines, trails, scaffolding, bigger trails and heroic perspectives until the synchronisation becomes a boisterous clamour.

8 This could also be said about the aesthetic of the pictures themselves which continues a specific ideological view of how to film moving bodies, which have a noticeable connection to the past when they "show tanned, muscled building labourers [...] at their hard day's work" (H. Goldschmidt, "Erster DEFA-Kulturfilm der Produktion Sachsen", in: *Tägliche Rundschau*, Berlin, 19-12-1946), which was quite similar to the pictures and the "view by which the German *Kulturfilm* always saw the German worker" (Günter Jordan, "Die frühen Jahre 1946 bis 1952", in: Jordan, Günter/Schenk, Ralf (ed.), *Schwarzweiß und Farbe. DEFA-Dokumentarfilme 1946-1992*, Berlin 1996, 15-47, 20). Only 8 years ago, Groschopp, the director of *Dresden*, was one of several cameramen of Leni Riefenstahl's *Olympia* (1938). The staging, the camera angle and the heroic views of the men show a continuity, which was only negated by the bodies themselves.

At the end of the film, we are shown the city's prosperity and liveliness, and rebuilt bridges, over which people are able to march as a united socialist subject, "with waving flags [...] compiling a performance that is symbolical for progress"[9]. These processes of assembling, connecting, synchronization, conformation are symbols aimed at the production of a political system. Groschopp celebrates this with a parade and grateful youths greeting the bridge builders. The bridge as a symbol of unity enters into an alliance with marching songs, with the homeland and a laughing and satisfied people.[10] After these events, the film produces a naive happy ending staging creative labour for the self-transformation and self-creation of individuals, ultimately attempting to carrying out a Marxist idea: the construction is for the constructors.[11]

Such a story of a building land is very deeply connected to the ideological aim of creating self-transforming protagonists in matters of economic and human progress, having as an important instrument, the poetry of socialism and concrete buildings. It becomes clear that building up, and especially rebuilding, was much more than merely a technical act. It described the substance of an ideology of progress that was aimed to end up in the merge of communism. The so-called "House of Socialism" and the "constructors of the future" had the leading role in the use of language in the Marxist theory as well as in literature, film and other forms of art. As a result of the increased use of such clichés, not only very typical rhetorical structures originated, but also a completely original concrete aesthetic arose, finally developing into a genre of its own.

This is how the building land became a self-description for the state. We are dealing with construction sites as well as with symbols that were staged in DEFA films and in other media. These media products are products of political legitimisation. The construction site was staged, glorified and also, of course, exploited: "it becomes a semaphore for transporting the symbol of upheaval to provide a projective mirror in the face of human self-transformation"[12]. Building sites are therefore a production site of ideology and political identification, so that they themselves produce meaning.

3. DISPLACEMENTS

Construction sites were always a place of accounting for the past, which not only means that they were a scene where memory and displacement meet. By showing the rebuilding process of post-war Germany, a new society could be established in contrast to the past. However, the building workers could only use the material of the old, so rebuilding always meant a recycling of the past, even if the DEFA films tried to rearrange it as means to create a new beginning. This creates a state of uncertainty, which dealt

9 Jordan, "Die frühen Jahre" (fn. 8), 20.

10 This bridge over the river Elbe is called *Brücke der Einheit* ("Bridge of Unity").

11 E.g. Brosig, *Es ist ein Experiment* (fn. 2), 186.

12 Ibid., 189.

with the impossible imagining of a completely different status. That is the *point* where the history of the construction site movies and the one of the GDR can be seen as a network of production, ideology and also representation operating within a staged reality.[13] After the first films of the early post-war years show modes of assembling material and individuals in order to build up human relations in a new society, the following years still concentrated on the process of rebuilding, but operated rather with different modes of time, obeying concepts of memory and oblivion. Like the films of the 1940s and early 1950s the construction site became the landscape for creating people and state. Nonetheless, in this territorial, architectural, political and least human development and re-creation, this transition of base and superstructure (especially in Marx' sense) leads at least to building resistance, if this corresponds to the negotiation of continuity and a repression of historical truth. Assembling until that moment did not mean something new, because there did not exist any new material, although, in contrast, change becomes clearly noticeable in a new aesthetic of architecture that wanted to demonstrate new ideals of power.

One of the first big and important building projects in the early 1950s was the construction of the Stalinallee in Berlin. It was more than just rebuilding a destroyed post-war area, rather being a symbolically overloaded "prestige" project that included ideological representations of the state's power and socialist ideals. Therefore the building of a new city block represents the power to construct something new out of ruins in combination with the power of the working class and a completely new start by breaking with the past. This ideology was also embedded in the architecture. The development design according to plans of Hermann Henselmann was quite similar to Moscow or Warsaw, based on a deterministic guideline of socialist urban planning. The so-called "socialist city" wanted a radical refutation of its history as a densely built city from the late nineteenth century. Transferring the concept of the socialist idea to an ideal city included the concept of overcoming fragmentation and socio-spatial segregation and therefore was influenced by ideological, political and social reform motives. After merging things and materials into one and portraying the labour force as a single working machine, the private and the living conditions had to assemble in a form of unity, all embedded in concrete. The new *Plattenbau* (a house made out of prefabricated concrete slabs) could be seen as simulations of the ideal GDR in miniature. The new building played a part in the big machine of the socialist city, divided into specific homogeneous housing units, composed out of replaceable, repeatable and reorganized mass-elements.[14]

A film by Kurt Maetzig was shot on the construction site of the Stalinallee in 1951 and mirrors the embedded ideals of that prestige project as well as the struggling issues of the protagonists working at this place. *Roman einer jungen Ehe* ("Novel of a Young Marriage", Maetzig, 1951) shows this site as a "project of the century", which brings technical

13 Ibid., 194.

14 That, so far, continued the concepts already shown in the here so called first period of construction sites in GDR films.

knowledge, material and individuals together in celebrating the power of the working class, who has given up its individualistic aims for the sake of the community [fig. 2]. The film has a married couple, Agnes and Jochen, as its main focus, both young actors who meet and fall in love in Berlin. Agnes then begins acting in East German films, which Jochen dismisses as propaganda, while he is working in the western part, thus leading to their mutual alienation. As the plot develops, he slowly becomes conscious of the ongoing rehabilitation of former influential Nazis have been in the western part of the city, whereas Agnes becomes ever more involved in the working class movement and spends most of her time as a volunteer at the construction site in the Stalinallee. Jochen's realisation of his being wrong about Agnes' stance leads to their reconciliation and living together in one of the new modern apartments of the Stalinallee's buldings. Although Maetzigs film could be seen as an ideal example of the first optimistic period, the fact that it shows GDR construction sites as a setting for demonstrating the building up of a new society also shows the conflict coming from the rehabilitation of Nazis in the western segments. This creates something completely new within the new GDR society. After assembling and uniting, this *operation of displacement* proceeds with the past and could be understood as the second main character of construction sites in GDR cinema.

A few years later, *Fräulein Schmetterling* ("Miss Butterfly", Barthel, 1966), which is also partially set at the Stalinallee but in the second building period in 1965, was a completely different film. Monstrously growing *Plattenbau* buildings start to displace and repress the remains of the traditional structure of the city and create a new arrangement for living [fig. 3].[15] Since the emtpy post-war rubble land was re-constructed, they started to de-construct and to demolish the old city districts around to produce new space

Fig. 2 *Roman einer jungen Ehe* (Screenshot, GDR, 1951)

15 The film was forbidden before it was finished. Today there remains only a working cut that was premiered as a reconstructed fragment in Berlin in 2005.

continuing their idea of a socialist city. Sure enough, that was an important turn: even the building land was always a place to replace the old with the new. This sense of permanent renewal through continuous progress in the GDR is conveyed via representations of repression by establishing the socialism construction to replace the past. Barthels film does not only show this, it also criticises this route to modernism. The theme of this film is no longer about a new society constructed alongside buildings and bridges. It may be a natural fact, that creating something new will always be connected with the cancellation of the past, or at least with the process of renewing the past's legacies. However, unlike Groschopp's film – which portrayed the rebuilding process emanating out of the ruins and remains in *Dresden* as a reconstruction, thereby attempting to "re-start" continuity by synthetically fusing, assembling and synchronizing the building-up of old bridges and houses between the ruins – the construction site of *Fräulein Schmetterling* is now overwhelming its immediate environment, taking over, displacing all things past.

This creates radically different perspectives of the construction sites inclusively in a broader, more general sense. Construction sites do not materialise out of a vacuum. Besides assembling, the process of *displacement* is a second focus which can be described more in detail using a method that helps us understand the psychological framework and its effect on construction sites (especially in a state like the GDR).

The site offered in GDR films can be described as a type of Freudian repression. According to Freud, the construction site represents a reorientation of presence and absence in the layers of the old and the new. The creation of something new often begins with the process of producing space. Thus the story of construction sites in the movies, and especially in the GDR films, often starts with demolition, blasting or decay. Between this process of letting the old vanish and the formation of the new, there are the somewhat technical

Fig. 3 *Fräulein Schmetterling* (Screenshot, GDR, 1966)

cultural operations of shifting, rearrangement and division. When work is done, it is not only the construction site that has vanished, but also the *space* which was previously there, as well as the objects and crucially the memory about everything that happened in this very place. The disassembly and deconstruction of entire old town districts should produce space for standardised buildings, but with "proletarian" luxury. An opposing method to the classic beginning in rubble films is a dichotomous segmentation between the material of the old and the new. The vanishing process of the old in the East German cities corresponds with the machinations of memory and the aim of "wanting-to-forget" or those of repression.

Following Sigmund Freud's concept of repression, everything starts through establishing a fundamental contraposition (such as consciousness and unconsciousness).[16] A primordial repression implodes the entirety into different parts. In the context of late wartime Germany, it could be seen in the destructive effect of the allied bombardment as well as the massive blasting actions. The second repression, the *actual* one, shifts the remains away to prevent them from being visible, therefore initiating the demolition of the images of memory. The process of repression wants "the past to be past", generating "a state of an ahistorical abeyance"[17]. Instead of the repressed, according to Freud, a replacement creates something new in order to establish another assembly of interrelations, which is the construction site itself with its own structures and operations. At this point it is worth pointing out that repression does not necessarily mean oblivion, but rather a pervasive environment based on a "not-wanting-to-remember". In that work of repression this ambivalence also forms the place where the repressed succeed in returning[18]. This operation, eventually, always leaves residual effects, so that the return of repression could manifest itself as a failure of the very politics of repression.[19]

4. CONTINUITY/DISCONTINUITY

"You get older. What do you think?" Li asked. For more than two years, car mechanic Al and midwife Li have been married, but since their wedding their life has stagnated. Disappointed and afraid of "missing out on something", they simply decide to end their marriage. The film then goes on to show a six month reflection period under judicial order until their divorce comes into effect. Since Al has vacation anyway, he uses his time in his own way: he strolls through the city. The situation of the generation born after the Second World War is one of lingering between past and future, a situation not of their own making that nevertheless forces and moulds them,

16 E.g. Freud, Sigmund, "Die Verdrängung", in: Ibid. (ed.), *Studienausgabe*, Vol. 3: *Psychologie des Unbewussten*, Frankfurt a.M. 1975, 103-188.

17 Götz Grossklaus, "Das zerstörte Gesicht der Städte", in: Böhn, Andreas/Mielke, Christine (ed.): *Die zerstörte Stadt. Mediale Repräsentationen urbaner Räume von Troja bis SimCity*, Bielefeld 2007, 101-124, 119.

18 Freud, "Die Verdrängung" (fn.16), 117.

19 That is the moment the dreamwork starts - in Freud's sense. If the repression fails, the remains return as a symptom. Odds and ends, rests and fragments are the material artifacts of the failed repression presented as a "dream of the things".

aligning them to the course of history. Jürgen Böttcher is a visual artist and documentary filmmaker. His *Jahrgang 45* ("Born in '45", Böttcher 1966) is a poetic yet realistic film that observes the first generation after the war as a portrait about rites of passage.[20] It explores both the present and a *possible* future by following Al on his strolls through the backyards of Berlin, its semi-finished constructions and rubble hills. The film is not a planned search, but a tentative exploration of possible perspectives and expectations of everyday life. It is almost a resigned future portrait that tries to forget all that has happened:

> In Böttcher's film the protagonists remain fundamentally disconnected from the memories of the war-time past that, in other films, are repeatedly invoked in order to legitimize and bolster the historical mission of the GDR[21].

Nobody wants to talk about the war experiences, not the protagonist's parents, not even museums, only the remaining rubble which constantly calls out for remembering, yet going coolly unnoticed by the protagonists:

> For Al memories of the fascist past – memories that he seems unable to tap into – are not sufficient to dispel the sense of existentialist despair with which he is beset.[22]

The film does not allow for any story, neither history nor historicity. The history of the generation post-1945 does not want to be named, just like the history of the couple is not shown – even their real reason for separation remains hidden. Only serial non-plots are left on display. They are used by the director in an attempt to characterise the film's vacant time as an attack against a lack of temporal coherence and the destruction of continuity by both state and society. The film strings scenes together, adding aimless actions without any purposeful interrelation. This kind of filmmaking follows the modernism of the early 1960s as seen in the western countries at that time, characterized by a dissolution of plot continuity and focusing on visual settings and situations.[23] Böttcher does not want to show a plot, rather an attitude presenting Al's longing for life in pictures. In a metaphorical way, there is always a need for construction work as an analytic method to process this unincorporated system by montage. At the end, it is only the gaze which could connect the different elements into one continuity or context.[24] This construction of images with empty plots and activities demon-

---

20 Shot in summer of 1966, the work for *Jahrgang 45* was prematurely terminated after complaints of the *Hauptverwaltung Film* of the Cultural Ministry of the GDR. Jürgen Böttcher continues working till the end of GDR as a documentary filmmaker, but eventually never realized a feature film. The remaining rough cut of *Jahrgang '45* could only be finished after 1989, premiering in 1990.

21 Allan, Sean, "The Love-lives of Others. Reconstructing German National Identity in Post-war and Post-unification Cinema", in: Passerini, Luisa/Diehl, Karen/Labanyi, Jo (ed.), *Europe and Love in Cinema*, Bristol/Chicago 2012, 153-170, 159.

22 Ibid., 161.

23 Deleuze, Gilles, *L'image-tempes*, Paris 1986; Fahle, Oliver, *Bilder der Zweiten Moderne*, Weimar 2005.

24 One scene might make that quiet clear: Al and his friends lounge around in front of the ruin of the well known *Gendarmenmarkt* in Berlin, lying on the stairs and watching the arrival of two buses of western tourists. Both sides interact only through a strange observation ceremony. Some tourists watch

strates the obvious need of getting connected by the operations of the film. The open ends readable through the technique of montage, discontinuity and the visible cuts or even *jump cuts* are showing a site of broken operations. Here the plot has been denied and what is presented instead is a picture instead of history, a situation about which Gilles Deleuze notes: "It is when history is broken that time presents itself in its pure state and deploys all the power of its non-chronological dimensions"[25]. Longing for continuity, only the construction site could merge things together, which is demanded by the film itself.

In Böttcher's film the construction site is not only the end of an empty interruption, but a beginning, a subsection and a replacement of the future. Only it is the short scenes that occupy the central position in this film. Al and some of his friends visit an apartment construction site in the *Plattenbau* outskirts of Berlin, the home-to-be of a young couple, barren vista of naked walls, open cable shafts and glassless windows. Suddenly, they all start to make plans: here will be the kitchen, there we will place the sofa and here you can sleep with a clear view of the stars at night. The future starts to appear. Whereas they roamed through the city earlier, they are now walking through this bizarre bare brickwork full of the not-yet-to-be future.

A few scenes later we see Al and Li on a motorbike. With a long lasting shot of the motorbike ride to the outskirts of Berlin, they gently go past the city, its industrial areas and shacks, also driving through natural landscapes with meadows and hills. In the moment they both stop at their destination, Li starts to look over the landscape. She focuses on one of the rubble strewn hills, a symbol for the accumulated past [fig. 4].

Fig. 4 *Jahrgang 45* (Screenshot, GDR, 1965)

the youths through the secure windows of the bus, others observe them from the outside with help of photo cameras. Al and his friends stand up setting them against the gazes. This process of setting gazes and making pictures substantiates the perspective of the film, the perspective of observing in fragments or details only, an incidental, spacial-temporal ensemble (Fahle, *Bilder der Zweiten Moderne* (fn. 23), 46).

25 Marrati, Paola, *Gilles Deleuze. Cinema and Philosophy*, Baltimore 2008, 78.

Since the late 1940s further building remnants were compiled and transported to the peripheral areas on the outskirts of the city for pouring and filling an assortment of hills after controlled disposal sites for operation of landfills had achieved their capacity. Millions of cubic meters were offloaded until the 1970s with the material remnants of tens of thousands of destroyed buildings. Over the years, some of them became the highest hills in the region of Berlin. Things decay, become remains, but *cannot* be fully eliminated and finally accumulate and conglomerate, becoming piles of rubble. Deploying Michel Serres's concept, we could characterise this picture by the figure of the *tabula rasa*, which is described by its own absence. The emptiness is the beginning of everything new and at the same time the continuation of an endless circle: "That history will never end"[26]. Much like the setting of a *reset*, qualified by Markus Krajewksi, who writes:

> [T]he tabula rasa [...] designated not least the impossibility of making everything forgotten. Of course, superficially it is a redemption, setting the counter to zero, a re-set of the present situation[27].

However, because resets always mean working with the setting of the residuals and remains as witnesses of the past, the *tabula rasa* never stays empty, so that "the remains are the motor of history, that follows after this non-remaining state"[28]. Here, the rubble hills pitted outside of the city, remaining a history that wants to be forgotten by the GDR, but on the other hand seems to be the material ground for building land. This picture is not only representative for the status of the GDR in dealing with continuity and discontinuity through the operations of assembling and repression, the camera views over the rubble hills also incorporate the method of the film itself, as a very self-referential picture connected with an image of bare brickwork. Li looks from the empty fallow land to the already bare brickwork of the *Plattenbau* estates, the camera shot starting at the rubble-strewn hills. That shot tells the whole story. The camera follows her gaze by alternating between her point-of-view and the filming of her eyes for connecting the movement of the shot and the gaze. The pan shot of the camera could be seen as a montage of layered time as it is connected to the rubble on one side and the *Plattenbau* on the other. Between both is fallow land, represented by Al's wanderings. By following that panned shot, the film envisions the linking between past and future as the only opportunity for reconciliation.This remaining framework of fragments and pieces of the past between gaps and openings creates a new arrangement. These are the first elements of everything new. That is how *Jahrgang 45* operates, too:

> Indeed what is so provocative about *Jahrgang 45* is its relentless questioning of what, in the absence of a theological view of the past and the future, might constitute a meaning-

26 Serres, Michel, *Der Parasit*, Berlin 1987, 278.

27 Krajewski, Markus, "Tabula Rasa. Der Rest stiftet den Neuanfang", in: Thuns, Barbara/Weiberger, Annette (ed.), *Was übrig bleibt. Von Resten, Residuen und Relikten*, Berlin 2009, 23-34, 27.

28 Serres, *Der Parasit* (fn. 26), 277.

ful existence in the here and now of the GDR in the mid-1960s, and it is in this spirit that Al resolves to recommit to his relationship with Li.[29]

While at the beginning *Jahrgang 45* tried to negotiate the methods of continuity, now it becomes clear at the end that there never was a reset and the "new" was in fact always corresponding to the past. That is why the reunion of the separated couple could only start at the end, when they were struggling to overcome the past, while the continuity becomes visible as the protagonist looks from one place to another, therefore connecting the past and the future in a long-panned shot.

Here, all things come together. Consequently, the last shot shows Li gazing directly into the camera, corresponding to the shot at the film's beginning. This, finally, is the instance of coherence the film was longing for from its beginning. After Al and Li searched in vain for connections (and therefore did not find any plot or history), a new perspective organizes the linkage from beginning to end, a connection which offers hope, both for the future of the couple and the generation of 1945. Thus, *Jahrgang 45* explores those generation's problems by the necessity to reflect upon the history, with which that they are forced to live.

4. THE END OF HISTORY

After we followed the representations of construction sites in GDR films from assembling to repression, we finally come to the symptom they diagnosed: the attempt of disconnecting from history creates a gap. However elaborately imagined, there never was a reset. Staging a construct of discontinuity only gazes into this gap between the past and the socialist ideal of the future. Again, it is all about the impossibility of making everything "forgotten". Because reset always means working with the setting of the remains as witnesses of the past, history always was the material ground for their building of land.

Trying to construe the construction site as a gap, searching for connections and starting to analyze construction work as the work of repression, we should also diagnose its symptom within the framework of Freud. The transformation of disassembling and reconstruction into displacement and repression can be traced in both the material and the structural. In the meaning of Freudian repression, the return of the repressed is a mistake of the unconscious, a symptom of repression itself. In the moment Li started to gaze from the rubble hills to the *Plattenbau* estates, history was visible again, continuity displacing discontinuity. In this sense, the methods of assembling, displacement and compression while building the material for a socialist dream means the imagination of constructing. The construction site isn't the dream, but a place where the dream can appear to close the gap, marks the transition between past and future and is moment of union, place of possibility and open entity.

29 Allan, "The Love-lives of Others" (fn. 21), 161.

This brief synopsis of the construction site in GDR film follows a story of failures and explains how this specific representation of a common place came to its end. Neglected self-realization, non-processed history, foreign objects and people in masses became building materials for the ideological concept of the GDR. However, with the passing years resistance against this political practice of the construction sites started to grow and increasingly films portrayed a critical view of decaying buildings, of senseless demolition and *Plattenbau* estates. Therefore, the radical renunciation of the past as well as the rigorous rebuilding of unified building estates became object of criticism. In order to juxtapose the opposite and its counterpart, the films often place the old and the new, the motifs of the decaying past next to the utopian creation. In the continuing cycle of urban construction, decay, demolition and reconstruction, modernity is represented as both liberation and total destruction.[30]

The progressive character of the construction sites in earlier years is now replaced by its emptiness and desolation, where there is no action and nothing progresses. Material shortages and the inability to finish major construction sites represented an excessive increase in the stagnation of fallow land. Since the 1970s, construction sites in DEFA films are pictures of stand-still. In a number of films from the 1970s and 1980s, there is an obvious increase in scenes about enduring abandoned construction sites. In these films, nonetheless, it is not the construction site itself anymore, which receive the main focus, much more interesting become the places in between, such as at the edges of concrete structures, between new buildings and ruined remains of old ones, in amorphous gaps. Often, fallow land is the only place where anything of a social nature happens. There are reasons for this choice of stage: it's the last space remaining undetermined, full of possible uses and perhaps freedom. Only the fallow land could offer a source for the imagination. In this context lies another important shift which classifies the last period of construction sites in DEFA films: in the moment when the idea and the utopia of the *Plattenbau* fails, the space represented by the fallow land becomes the latest (new) escape.

That is why *Die Architekten*, as mentioned at the beginning of this paper, was the least possible film in the GDR. During the entire plot, the vision and plans of the construction site remain unfulfilled. The public administration initiates a massive intervention to guarantee the start of construction, but the architecture then is of a uniform manner that had already been built many times in many other places. This shows how the construction site itself represents the psychological status of the GDR. Beginning with an ideological (and pathetic) staging it evolves into a setting for the failure of the country's own utopia. The standstill, the unfinished alongside unchanging ideological concepts represented in the late DEFA film by the fallow land show the failure of the GDR in the material and the imaginary world. The film is both the last film and the last construction site

30 Other well known examples are *Die Legende von Paul und Paula* ("The Legend of Paul and Paula", Carow, 1973) or *Unser kurzes Leben* ("Our Short Life", Warneke, 1981). The representation of demolition and construction as modes of replacing the past are consequently situated alongside an understanding of architecture as utopia put into practice, which means making the absent present.

of the GDR to be looked at. The moment of the end of the construction of the material and imaginary marks the end of the belief in the construct of socialism. The construction site, which, as has been demonstrated, is thus not only used to trigger oblivion, it is a point of convergence in the history of the GDR, representing its ideology and utopian ideals, as well as its implosion.

LIST OF FIGURES

PART 2:
THE CITY UNDER CONSTRUCTION

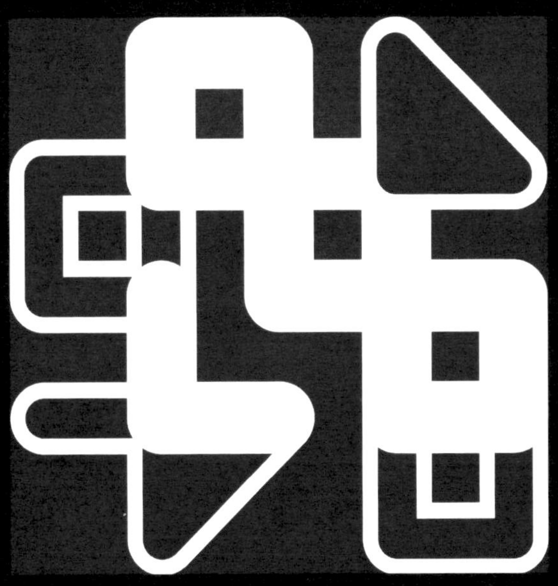

NEW NEW YORK, A CATCH-ALL NARRATIVE FOR A NEW METROPOLITAN IDENTITY

Patrick Leitner

Between the turn of the twentieth century and the beginning of World War One different administrative, social, technical, economic, and cultural evolutions came together in New York. These included the creation of the Greater New York City in 1898, resulting in an instant increase in the city's population from 1.850.000 to 3.500.000 inhabitants and in a roughly five times increase in the city's area; the introduction of electric traction which meant that the heart of Manhattan could be reached more easily through tunnels; the boom of American capitalism, making available unprecedented amounts of money for building; and the more general goal of becoming a world city of London's and Paris' status.[1] As a consequence, New York City was the stage on which huge urban transformations took place. Infrastructure and transport facilities, office and apartment buildings, public buildings and monuments, among others, changed at a rapid pace over the course of ten to 15 years. It was a city changing its scale, functioning, and identity. The radicalism of change was at that time only comparable to the modernization of Paris carried out during the Second Empire (1851-1870, under Baron Haussmann) and continuing through the French Third Republic (from 1870 onwards). It was a transformation that New York's cultural elites continued to look at with great envy.[2]

What makes the case of New York around 1900 so compelling is the fact that the diverse physical alterations of the city were accompanied by a discursive construction of reality. From 1902 on, the principal New York periodicals of general interest contributed to this: *The American Magazine, The Century Magazine, Cosmopolitan, Harper's Monthly Magazine, The Independent, Munsey's Magazine, The Outlook, Scribner's Magazine*, and *The World's Work*. Several observers gathered the on-going changes under the notion of "New New York" which appeared as the title for some of their articles and also for one important monograph. While this notion seemed to imply the existence of a comprehensive plan or of concerted action to actually renew the city in a coherent way, I first aim to show that it was mainly if not exclusively used as a performative tool designed to produce a smokescreen narrative of legitimization and unification for the actual and mostly uncoordinated physical world building process. Secondly, I aim to illustrate that while the physical and discursive building processes went well beyond a single project and its related construction site, some specific

1 For a study of the late nineteenth century New York discourse on its status, see Leitner, Patrick, "Un ménage urbain à trois, ou l'ambition mondiale de New York", in: Cohen, Jean-Louis/Arnold, Dana (ed.), *Paris Londres*, Gollion (Switzerland) 2015, 385-433.

2 The complex relationship between Paris and New York in the decades around 1900 is analyzed in Leitner, Patrick, *Entre Paris et New York, dynamiques d'échange pour transformer les métropoles, 1858-1926*, unpublished PhD thesis, Université Paris-8 2009.

objects, namely Grand Central Station, played a role more important than most in the definition of elements that were applied in the New New York discourse on the entire city. Thirdly, I argue that the notion of New New York served as the foundation of the ongoing discourse of ambition and competition, both at a local and inter-metropolitan level.[3]

New New York is also exemplary for showing in what ways the imaginary world is built before, during, as well as after the material world. While the building of the material world happens in a specific timeframe – it only occurs when it actually materialises –, the building of the imaginary world happens continuously, though more or less intensely. The process generally starts with the very first critique of a material situation, is followed by the first idea of physical alteration and its evolution in a series of projects, continues through the construction work itself and covers the entire life and even afterlife of the changing material world. In a way that precedes our contemporary – and nowadays mostly official – ways of communication on physical changes to the city, the building of the material world back in 1900 in New York was but a small part of the manifold and complex evolutions of the discursive world.

1. CRISIS

The first step in the narrative of New New York consisted of the outspoken acknowledgement of a crisis. New Yorkers themselves expressed in the last decades of the nineteenth century frequent claims of their city being architecturally provincial and in urban disorder.[4] From 1900 onwards, however, mainly the problem of traffic and public transport was highlighted. The expression of this crisis formed one of the founding elements of the imaginary world building process, an element indispensable for the ensuing action of physical infrastructural alteration.

Mainly two reasons were used to explain the problems: Firstly, the geographic situation of Manhattan, surrounded by deep rivers, which had caused the city's isolation and resulting congestion; and secondly, the inaction by the city to tackle the problem. The lack of action had resulted in a general backwardness, as noted in the journal *The World's Work* in 1903:

---

3 Ultimately, these arguments will offer a new perspective on the notion's range and scope that were much wider than what scholars usually have conceived. Although New New York does feature in the title of books, for example in Douglas Tallack's *New York Sights, Visualizing Old and New New York* (2005), or in the title of book chapters, for example in Angela M. Blake's *How New York Became American 1890-1924* (2006) and in Mary N. Woods' *Beyond the Architect's Eye: Photographs and the American Built Environment* (2009), or as a key element of the argument, for example in Wanda Corn's *The Great American Thing: Modern Art and National Identity, 1913-1935* (1999), the notion is mostly if not exclusively limited to mentioning what John C. van Dyke had presented in his 1909 book *New New York* and to what the photographer Alfred Stieglitz had captured in his famous photograph "Old and New New York" of 1910. As such, at least to our knowledge, no study of New York has offered a detailed analysis of New New York, and even less so an analysis of the notion itself as a discursive tool serving different if not opposing agendas.

4 See Stern, Robert A. M./Mellins, Thomas/Fishman, David, *New York 1880*, New York 1999, 9, 13, 19; and Kantor, Harvey A., *Modern Urban Planning in New York City. Origins and Evolution, 1890-1933*, unpublished PhD thesis, New York University 1971, 6-12.

> The population and traffic density in Manhattan [...] are far beyond those of any large city. [...] It is instructive to look back and discover that in Manhattan there have been no really great additions to the transit facilities within the past quarter of a century beyond the building of the Brooklyn Bridge and the elevated roads. With travel increasing 100 per cent every ten years and no additions to the track mileage worthy of mention, there was bound to come a time when the existing lines [...] would reach the limit of their capacity. That time has now come.[5]

However, the inaction affected not only the state of the general network but apparently also every single means of public transport. The horse-cars were being described as "antiquated" or "a survival from a bygone era", as in the captions of photographs published alongside the text which showed horse-drawn cars that provoked unacceptable slowness: "The slow horse-cars, the time-destroying ferry service [fig. 1] and the poor suburban car service have confined the population within the limits of a narrow circle [...]."[6] In a more retroactive look, *The American Magazine* noted in 1907: "Manhattan Island was crowded to the bursting point, but its rivers kept it in, for the ferries were quite insufficient as an outlet, aside from their inconvenience and the time they consumed in transit."[7]

Slowness was not the only consequence, though. Photographic images of extreme situations of rush hour were supposed to convincingly show the overcrowded and inconvenient conditions in which people travelled.

Fig. 1 The time-destroying ferry service.

5 Wheatly, W. W., "Transporting New York's Millions", in: *The World's Work*, May 1903, 3425.

6 Ibid., 3436.

7 Eaton, Walter Prichard, "Manhattan: An Island Outgrown", in: *American Magazine*, July 1907, 246.

These went as far forcing people to make risky trips by standing on the front and rear platforms of a street-car [fig. 2]. And even in the case of the elevated train or the train over the Brooklyn Bridge, where the devices themselves were not depicted as outdated or slow, "the mobs [had] literally [to] fight for conveyance during the rush hours".[8] Some of the lucky ones who did actually board the train would only find a spot by bridging the gap between two carriages. In order to give this problem a supposedly objective or even scientific proof, the crisis was also heightened with the help of statistics.[9]

As if the crowded cars and the statistics did not suffice, the "disorder and discomfort"[10] of New York transportation, as one journalist still called it in 1913, was underlined by the fact that the streets themselves were congested: by people coming out of big office buildings and, paradoxically, by the very same public transport street cars whose ever increasing number, lined up along Broadway, was supposed to counter the problem of overcrowded means of transport.

After having portrayed a seemingly inextricable situation, some articles however offered the prospect of a way out: "But the fourteen new tubes under the rivers will change all this."[11] This would make "certain that real rapid transit with ample accommodations will remove the radius of

Fig. 2 An overcrowded street-car.

8 Wheatly, "Transporting New York's Millions" (fn. 5), 3431.

9 Wilcox, Julius, "The New Transit System and the New New York", in: *The Independent*, 24 September 1903, 2285.

10 Marcosson, Isaac F., "The World's Greatest Traffic Problem", in: *Munsey's Magazine*, June 1913, 328.

11 Eaton, "Manhattan" (fn. 7), 246.

the [city] circle farther and farther into the country districts",[12] promising some relieve to the heart of the city. In general, as the same text of 1903 added reassuringly, "Greater New York is now on the verge of a gigantic transformation".[13]

2. WAR

The portrayal and description of this "gigantic transformation" by the same general public media outlets, as well as in official documents, also reveal the building of an imaginary world related to it. Three elements are here of particular importance because they were supposed to support the narrative of the gigantic transformation, the notion of the *complete* rebuilding of the city.

The first element is the constant mentioning of the huge if not unprecedented efforts that were being made. Not only in financial terms – "I do not know of any other city or any other country that can spend such an amount for transportation purposes"[14] – but especially in physical terms, since many observers underlined the fact that it was solid rock that needed to be removed for every single construction site. At the simultaneously happening construction sites of three train stations, "one sees hundreds of men busy in removing, with compressed air drills and dynamite, and huge, derrick-suspended steam shovels, millions of cubic yards of stone and silt".[15]

The second element consists of the chaos, the noise, inconvenience, and obstruction caused by these works which was used as proof of a complete rebuilding of the city, for example in the 1901 article "The Rebuilding of New York" in which the very first phrases were as follows:

> A little sunny-haired girl was looking through the dingy window pane of a down-town surface car in New York. On every side buildings were being torn down and the ugly framework of new structures was going up. Thirty-fourth Street stood up like a ridge out of a great hole of excavations. [...] The dinning of hammers, the shouts of men, the rumbling report of a blast, the wheezing of drills, and the litter and obstruction everywhere gave the impression of a great quarry and a mammoth workshop. 'I hope it will be a real nice city when it's all done,' said the little girl. Such is the impression that New York just now makes on anybody – a torn-up place, a city in process of complete rebuilding.[16]

With regard to the excavation for the Pennsylvania Station [fig. 3], one could read:

---

12 Wheatly, "Transporting New York's Millions" (fn. 5), 3436.

13 Ibid., 3426.

14 Rice, George S., "Rapid Transit in New York City, paper presented to the Municipal Engineers of the City of New York, September 30, 1903", in: *The Municipal Engineers of The City of New York, Proceedings for 1903*, New York 1904, 71.

15 Bruère, Robert W., "The Gates of New York", in: *The Outlook*, 27 April 1907, 928.

16 Cunniff, M. G./Goodrich, Arthur, "The Rebuilding of New York", in: *The World's Work*, December 1901, 1485.

> The signs [of the building of the tunnels under the river] at the terminal, however, are numerous enough, for at present the station is a yawning hole half a mile long and two blocks wide like a gigantic earthquake freak, with three avenues, an elevated structure, sewers, water mains, gas pipes and the like straddling across it on steel stilts, performing their normal functions.[17]

Finally, metaphors of war were supposed to help understanding even better the scale of the works, the size of the efforts, and the "the blasting that had been done at all hours of the day and the night".[18] These metaphors, in turn, might explain why barely any workers can be seen on the photographs, merely battlefields [fig. 4 & 5]:

> Brick and timber and stone filled the streets along with the big steel girders for the new constructions. This went on up-town and down-town till there is a wide line of desolation and destruction from the Battery to the Bronx. It is as if a cyclone had expended its force upon the city or as if a hostile fleet had bombarded the island.[19]

That idea was relatively widespread since it resurfaced in 1902 in an article of *The Century Magazine*:

> One might almost fancy that the town had been bombarded by a hostile fleet, such rents and gashes appear everywhere in the solid masonry, ranging from the width of a single building to that of a whole block front, as where the new East River bridge has made footroom for itself on the Manhattan shore. The very spine of the island has been split by dynamite in preparing the way for rapid transit; and where excavations are being made in preparation for certain new buildings, it looks as if lyddite shells had exploded, ripping up tons upon tons of bed-rock and gravel.[20]

Fig. 3 Excavating the Pennsylvania Station (left).
Fig. 4 Subway works at Union Square (middle).
Fig. 5 Scenes of "bombardment" in front of Grand Central Station (right).

17 Eaton, "Manhattan" (fn. 7), 249.

18 Rice, "Rapid Transit in New York" (fn. 14), 67.

19 Cunniff/Goodrich, "Rebuilding of New York" (fn. 16), 1486.

20 Blackshaw, Randall, "The New New York", in: *The Century Magazine*, August 1902, 493.

3. HARMONY

It is interesting to note that the imaginary worlds built out of the description of the actual construction site were often put in huge contrasts, if not opposites, with the future. These contrasts were constructed by observers in order to compensate and sometimes overcompensate the violence of the situation with the help of an exaggeratedly positive description of the improvement to come. For example, as mentioned earlier, the little blond girl facing the chaos and yet dreaming of a "real nice city" in which she would perfectly fit in; or, the war-like "desolation and destruction" of the subway works described in an article whose soft tone illustrations show New York as a civilized and harmonious place of neo-classical architectural monuments but with no apparent existence of modern transportation [fig. 6-9]. Similar types of images, but of Grand Central Station, would appear a decade later in its opening year 1913. Here, softness and quiet dignity of drawings were also supposed to express the spirit of the civic centre "for the public good", contrasted strongly with the violence of its construction site. Regarding the railroads in general, there was another contrast to be pointed out. Gas, smoke, steam and danger would make place for the future, noiseless, clean, safe, and almost unnoticed passage of electric trains, as described here in 1907:

> In a short time – probably three years at most – four railroad systems will bring their enormous traffic into the very heart of Manhattan Island under rivers and streets and avenues, without a puff of smoke or a sound of steam. Underground, in silence and clean air, they will come and so depart again [...].[21]

The ultimate goal of the sacrifices, however, was best described in the following terms of contrast:

> Never in the history of the world has a whole city so tortured itself in the eager desire for comfort, luxury, beauty. For all is but in anticipation of a stupendous transformation that will make Manhattan unequaled as a civic monument to modernity.[22]

These examples of using contrasts show that the narrative of compensation had to resort, necessarily, to the anticipated results. As the last quote illus-

Fig. 6-9 A civilized, calm place.

21 Eaton, "Manhattan" (fn. 7), 245.

22 Arnett, Frank S., "The Evolution of Manhattan", in: *Munsey's Magazine*, November 1902, 178.

trates, different themes – in this case, comfort, luxury, beauty, modernity – associated to the so-imagined future improvements and identity strengthened the narrative of improvement.

The first anticipation was the one of future speedy access to the heart of the city, even from the outer boroughs and across the Hudson River. While some mentioned the fact that it "will take only 5 minutes to get from Jersey City to Wall Street",[23] others underlined the fact that "[by] 1904, a worker at the crowded end of Manhattan Island will be able to reach a distant home with speed and comfort"[24] and, also, that

> dwellers on Long Island who now must wait for ferries, and fume in crawling surface cars, will be picked up as they reach the steam terminal, shot under the river, lifted up to the Subway level at Park Avenue and Thirty-third Street, and then bowled up or down town with touch-and-go rapidity.[25]

However, the general aim behind the need for speed was not only – or even, not so much – individual comfort but a far superior goal, i.e. the economy of the city:

> The big fact that stands out in all this is that New York wants to be able to reach its office just as quickly as can be. [...] Minutes are great big dollars in this nerve-consuming town. That is why New York is pouring out such fabulous riches for its transportation, building homes and offices upward and downward and connecting them by the most complete system of surface and elevated railroads, tunnels and bridges in the world.[26]

Chronologically the next theme, the future beauty of the city, was much more based on personal interpretation but nevertheless a very important if not the most important one for the building the imaginary world. Regarding the tall buildings, one can read the following in 1905 in *Munsey's Magazine*:

> Already the annual increase in skyscrapers alone is counted in hundreds [...]. And why not? [...] They may add a Babylonian beauty to the city [...]. Yet when New York is covered by these [office buildings, hotels, and apartment buildings], when they have ceased to be the exception and have become the inevitable, they will have acquired the serenity of a natural law, and with a beauty titanesque in grandeur.[27]

One year later, *The World's Work* writes: "The buildings under way at the time this article is being written would in themselves make a city of no mean proportions. Better still, they would make a city of great architectural beauty."[28] The article concludes: "Thus do beauty and the ideal blossom from out our gross material prosperity."[29]

---

23 Eaton, "Manhattan" (fn. 7), 253.

24 Cunniff/Goodrich, "Rebuilding of New York" (fn. 16), 1491.

25 Ibid., 1506.

26 "What New York Pays To Get To Its Office", in: *The World's Work*, October 1902, 2680.

27 Saltus, Edgar, "The Colossal City", in: *Munsey's Magazine*, March 1905, 790.

28 Bowles, J. M., "A New New York", in: *The World's Work*, December 1906, 8301.

29 Bowles, "A New New York" (fn. 28), 8306.

4. BETTER STRUGGLES, STRONGER CITY

Nonetheless, the discourse on the rebuilding of New York included a quite different aspect, one of struggle at all levels as a sign of New York's vitality. Diametrically opposed to the discourse on future harmony, it continued in a sense the narrative on war-like destructions of New York. Indeed, according to some articles, quicker access to the city was not only a question of going to work and back home as quickly as possible, but foremost one of improving the conditions for the daily struggle of people against each other:

> Eventually Manhattan Island will be a great rampart of steel and stone, which each morning from every side, across bridges, through tunnels, by train, by boat, on foot, hundreds of thousands of men and women will storm, and at night will drift back to rise next day to renew the fight. For New York must always be the amphitheatre of struggle. It is because the struggle has become more swift and strong that New York is being rebuilt.[30]

This idea re-surfaced in 1903: "Freshened and strengthened by their closer contact with nature, [the workers as well as the business men] will come back to their work each morning to renew the struggle for existence or supremacy in the most intense city in the world. [...]"[31] And in 1905: "[With subway and tunnels] hordes of human beings will descend and scheme and fight and lie and die for gold."[32]

Interestingly, the question of struggle and survival – which, ironically, was not far removed from the war-like environment of the subway and train station works – was also applied to the buildings, implying that the material city, too, fought everyday for its existence [fig. 10]. Just as one paragraph heading in 1905 read "The Endless Process of Reconstruction",[33] the author of "The New New York" in *The Century Magazine* in 1902, in spite of publishing images of a harmonious, calm and settled city, explained the constant rebuilding process as the expression of a natural law:

> Reckless as all this seems, wasteful as some of it undoubtedly is, by far the greater part of destruction wrought has been commercially inevitable, and in accordance with a law of growth that involves the reconstruction of the city's central and more crowded quarters simultaneously with the pushing forward of its frontiers. A hundred thousand dollars must be sacrificed, if necessary, to provide for the advantageous investment of a million.[34]

Similarly, the homonymous article published in 1906 in *The Outlook* noted the "transformation upon transformation forced upon [the huge organism] by the necessities of growth".[35]

---

30 Cunniff/Goodrich, "Rebuilding of New York" (fn. 16), 1511.

31 Wheatly, "Transporting New York's Millions" (fn. 5), 3436.

32 Saltus, "The Colossal City" (fn. 27), 791-792.

33 Ibid., 790.

34 Blackshaw, "New New York" (fn. 20), 496.

35 Baxter, Sylvester, "The New New York", in: *The Outlook*, 23 June 1906, 410-411.

Fig. 10 The natural law of eternal rebuilding.

The scope of struggle was further enlarged to make the whole of New York compete on a global stage on which it was already presented at being at its center, for example in this passage of 1903: "Truly we shall have within a few years a Greater New York — greater in many ways than we could have foreseen twenty years ago, and perhaps the greatest centre of population and commerce the world has ever known."[36] And, again, in 1905: "[Manhattan] will be the business center not of a metropolis but of a nation, a great mart, the greatest in the world."[37]

Thus, just like the discourse on future beauty, the discourse on eternal struggle and unlimited ambition – and, therefore, necessary constant rebuilding – was explained with the idea of a natural law. The latter discourse however declared the natural law of struggles as something inherent to New York's nature, from people to building to the entire city. In return of that logic, labelling the city the greatest in the world implied, of course, that the struggling people in it were the greatest, too.

5. NEW NEW YORK

Nevertheless, as contradicting as these two discourses seemed to be – on future harmony on the one hand, and on endless struggles and reconstruction on the other –, they were all covered by the notion of 'New New York'. The four articles whose titles included the terms, published between 1902 and 1906 in the journals *The Century Magazine*, *The Independent*, *The Outlook*, and *The*

36 Wheatly, "Transporting New York's Millions" (fn. 5), 3436.

37 Saltus, "The Colossal City" (fn. 27), 791.

World's Work, applied it varyingly to these different aspects of the city.[38] Randall Blackshaw's "The New New York" and J.M. Bowles' "A New New York" focussed mainly on the future architectural (and mostly classical) beauty of the city, while also mentioning the beneficial recurrent reconstruction of the city according to the law of growth. As seen earlier, Sylvester Baxter had shared this last aspect. However, in his 1906 article "The New New York", Baxter illustrated even more strongly the ambiguous nature of the notion and the difficulty to fully embrace its realities. For although he briefly enumerated the manifold recent improvements of the city, ranging from new parks to cleaner streets, great bridges and tunnels while also including better schools, better buildings and better tenements creating "a better civilization all round",[39] his text concentrated on the most visible aspect, the new landscape the skyscrapers had created. Surprisingly, though, a detailed reading of the article revealed some regrets regarding their consequences. This seemed to be heightened by the illustrations of which a large majority showed new but low buildings like banks and churches [fig. 11-14], very similar in architectural appearance to those published in Blackshaw's article four years earlier. The fourth article, Julius Wilcox's "The New Transit System and the New New York" published in *The Independent*, concentrated on population statistics and questions of rapid transit. As for John Van Dyke's 1909 book *The New New York*, it emphasized yet another aspect: the interest for the city's fast-changing, strongly contrasted and unique landscape as an expression of its huge commerce, great power, tremendous vitality, and individual freedom.[40]

Thus, it becomes clear that the notion, precise as it sounded, was in fact just as vague as it was open and could therefore include even opposing agendas regarding the city's perceived state and desired future development. The desire to create there a coherent *discursive* world was also underlined by

Fig. 11-14 New low buildings in vertical New New York.

38 Blackshaw, "New New York" (fn. 20), 491-513; Wilcox, "New Transit System and New New York" (fn. 9), 2283-2286; Baxter, "The New New York" (fn. 35), 409-424; Bowles, "A New New York" (fn. 28), 8301-8306.

39 Baxter, "The New New York" (fn. 35), 424.

40 Van Dyke, John C., *The New New York*, New York 1909.

the fact that four texts out of these five mentioned *the* New New York – rather than *a* –, the definite article clearly emphasizing the specificity, if not proof of existence, of the object that was written about. Furthermore, in addition to the *tour de force* of including aspects of future harmony alongside aspects of eternal reconstruction and stark contrasts, employing the term "New New York" had also suggested some sort of comprehensive planning. Regarding the buildings, this was obviously not the case. But in order to dispel the idea that the absence of coordination between buildings would have a negative impact, "The New New York" article in *The Outlook* went as far as declaring that the buildings' "rampant individualism" was the basis for the unintended positive result. Seen from the bay, the overall image of the city, resulting "from the successive efforts of perhaps half a hundred different architects", was something "so colossally imposing" that "premeditated study" could not have achieved.[41] Regarding the works of transport facilities, however, the idea of a comprehensive scheme was at least presented as such, for instance in the title of the article "The New Transit System and the New New York". Nevertheless, these projects were in reality started individually and "without preconcerted action",[42] leading to calls for "a comprehensive plan".[43] Incidentally, an example of the individual nature of the works could be seen in an advert published in 1910 for the upcoming opening of the Pennsylvania Station. The accompanying map featured all the routes of the Pennsylvania Railroad while the network of the New York Central Railroad was glaringly absent.

Individualistic buildings formed an unintended collective result on the one hand, and supposedly coordinated networks formed independent entities on the other: all this – and much more – could be fitted into New New York, the only proper criterion was the relatively recent construction of the object in question.

6. TRAIN STATIONS

Nevertheless, the train stations did represent important elements in New New York's narrative. Pennsylvania Station, the Hudson Terminal, and Grand Central Station (or Terminal) were the three new stations being built in Manhattan in that period. Beyond all practical considerations and the shared feature of subterranean tracks, each one of those stations played a specific role in the narrative.

Pennsylvania Station was conceived as an exclusive train station built over underground tracks. Its building was accompanied by tales of a huge construction site and of future uninterrupted national railway through traffic from the mainland to Long Island. Furthermore, it was hailed for its architectural calmness and dignity, and for being a place devoid of any kind of business activity. It was thus part of the narrative of destruction and future harmony. As for the Hudson Terminal, it opened in 1909 as an entirely underground station, devoid of a proper station building and thus invisible

41 Baxter, "The New New York" (fn. 35), 415-416.

42 Cunniff/Goodrich, "Rebuilding of New York" (fn. 16), 1486.

43 Wheatly, "Transporting New York's Millions" (fn. 5), 3423.

from the outside. Because it was built over by four huge identical office buildings, it was (somewhat wrongly) called "the biggest office building in the world", "the largest building in the world", "a city in itself", with accounts of the number of people working there varying from 10.000 to 25.000. It was part of the narrative of access to the city and its new scale.

Finally, Grand Central Station opened in 1913 at the heart of a vastly reshaped neighbourhood. Since it ticked all the boxes of the New New York narrative, it was its most complete representative, and this beyond the sole group of train stations. The following quotes from a *Munsey's Magazine* article of 1911 illustrated virtually all of the elements: "Out of what has until lately appeared to the casual traveller to be a confused hole in the ground, criss-crossed by tracks and intermittently shaken by blasts of dynamite, the world's greatest railway terminal is rising in the very heart of New York City."[44] And further:

> The reclamation of the city blocks formerly devoted to unsightly railroad yards, and the harmonious architectural group plan by which a whole new civic center is to be evolved [fig. 15], [...] is perhaps the largest, and it promises to be the most successful, combination of the esthetic and the practical in city building yet planned in America.[45]

Other features contributing to Grand Central Station's importance were the five superimposed layers of railroad and rapid transit situated just in front of the station under 42nd Street, turning the spot into an easily reachable heart of the city. Additionally, the station included two superimposed layers of tracks, directly related to the city's rapid transit. Lastly, it was built over not only by a classical European-style monument – "for the public service" as it was said – but also, as was planned, by a "great new office building"[46] as a modern American monument to money and materialism [fig. 16]. The station was thus able to represent everything that was included – and seemingly contradictory – in

Fig. 15 An architectural group plan for Park Avenue (left).
Fig. 16 Project for an office building over Grand Central Station (right).

44 Thompson, Hugh, "The Greatest Railroad Terminal in the World", in: *Munsey's Magazine*, April 1911, 27.

45 Ibid., 38.

46 Bruère, "Gates of New York" (fn. 15), 928.

New New York: an easily accessible, civilized and harmonious city; and a business city emerging out of constant reconstruction and engineering wonders. The great variety of images published to illustrate Grand Central Station clearly emphasized these roles. Besides, the fact that Grand Central Station was the only one of the three stations under construction to be transformed and not newly built did further enhance its unique position. Due to this fact, its images enabled showing the three key moments: the unsatisfactory situation before construction, the promise of future harmony, and the actual construction site [fig. 17-19].

Fig. 17 The New York Central yards in 1900 (top).
Fig. 18 The New York Central yards under construction in 1907 (middle).
Fig. 19 A drawing of the projected lowered and covered New York Central yards (below).

7. POSITIONING

Supposed to build for New York a coherent discursive world out of diverging material elements, the New New York narrative was also meant to help redefine the position of New York in relation to other cities of similar or, seemingly, higher status. For, in the end, that was its key objective: to convince that everything was, or at least would be, new, splendid and authentic, and that it would be possible, and almost inevitable, to surpass New York's nearest rivals London and Paris in their respective domains of pre-eminence. Van Dyke's *New New York* illustrated this very clearly. From the outset, and repeatedly through the over 400 pages of the book, the author tried to make his point about New New York being not only a new New York grown out of its place and people but also a new type of city, unlike anything that had ever been built in Europe and, most importantly, not in the slightest sense inferior to neither the British nor the French capitals.

This comparative approach was recurrently used in various journals of that period. Many passages provided evidence for this, for example with regard to richness: "Never, at any time, in any form of society, [...] has civilization beheld greater lavishness than that which our metropolitan plutocracy displays. More ornate than the swirl of London, and more resplendent than that of Paris."[47] However, most passages referred to beauty and magnificence:

> Now that New York has grown to be so big and so prosperous — now that she is the most strenuous of the world's cities and almost the wealthiest — she has set her heart upon being the most beautiful as well. [...] If her guests do not praise her of their own accord, she lures them on with eager questions. 'How do I look? Did you ever see anything quite like me? Do you not really think that I am handsomer than London and more stylish than Paris?'[48]

While London was usually mentioned only with regard to size, wealth and population, and only exceptionally mentioned in this context, the usual and recurrent reference with regard to beauty and civilization was exclusively modern Paris as rebuilt by Napoleon III and the following French Third Republic. In *The Century Magazine*'s article "The New New York" of 1902, one could read: "But to-day a new New York is coming to birth which bids fair to vie, if not in historic interest, at least in magnificence and beauty, with even so splendid a capital as that of France."[49]

In such texts, New New York was thus clearly imagined as ambitious and insubordinate towards its supposed rivals. However, other texts using comparison as a means to position New York yet not based on the notion, were much more nuanced if not sceptical, as the following passages from the article "To Make New York Beautiful" from *Outlook* in 1902 showed:

> Clean streets, good government, thorough order, the best conditions of the civilized life of cities, must be secured in America by the co-operation of citizens. This is our problem, and it is a very much more difficult problem than it was under Napoleon III. [...]. The city is being marred

47 Saltus, "The Colossal City" (fn. 27), 793.

48 Casson, Herbert N., "New York, the City Beautiful", in: *Munsey's Magazine*, November 1907, 178.

49 Blackshaw, "New New York" (fn. 20), 493.

by great edifices going up in every direction which are simply iron cages, without form, grace, or beauty. They represent in iron and stone pure materialism [...]. It is impossible to secure in American cities the uniformity which exists in Paris, because Americans will not accept the restrictions which make the imposition of such uniformity possible; but it is possible to educate American public opinion to such a degree that no man will wish to build an ugly structure [...]."[50]

Others went as far as expressing a real regret for not being able to reach what Paris had taught them to appreciate:

[On Fifth avenue], beneath the sempiternal blue of the sky the effect is gracious. Were the avenue itself wider, it might ultimately suggest the Paris Elysian Fields — might, we say, yet never will, if only because of the cyclopean scarps of the buildings.[51]

While this desire to find in New York aspects of a monumental, dignified but also picturesque Paris was widely shared,[52] the simultaneously happening New New York discourse clearly allowed for a defiant position.

8. FROM AFAR AND FROM WITHIN

In so far as Paris was seen as New York's major benchmark, what was known *there* of the New New York? How was all this New York upheaval looked at from the French capital? All the new material elements of New York were presented and analysed at the same time in Paris, except for the classical 'European style' monuments that were of no interest to the French. However, compared to American ones, French sources showed a huge difference as not a single one of the many articles on New York enumerated all or even some of the different material changes going on there. The various urban elements of New York were discussed in distinct articles and even in distinct journals. Journals of general interest, in particular *L'Illustration*, expressed an increasing fascination with the skyscrapers, the formerly "monstrous objects" that had turned into something of "beauty and elegance".[53] However, neither the infrastructural feats, nor the train stations, nor the quest for harmony, nor the eternal reconstruction were mentioned in the same publications. On the other hand, the journals with a more scientific or technical approach like *La Nature* or *Le Génie civil* presented New York's infrastructural projects, but not the general landscape of the city. It is in these sources that one could find many accounts of train stations among which Grand Central was the most discussed and hailed as the "master piece" of its kind.[54] And while the cityscape of big

50 "To Make New York Beautiful", in: *The Outlook*, 23 August 1902, 1006.

51 Saltus, "The Colossal City" (fn. 27), 789.

52 Leitner, *Entre Paris et New York* (fn. 2).

53 "Le quartier des immeubles géants de New York vu des échafaudages de celui qui les surpassera tous", in: *L'Illustration*, 3 August 1912, 80-81. For an overview on the shifting topics on and the increasing fascination of high buildings in New York seen from France, see Leitner, Patrick, "Lost for Words. How the Architectural Image Became a Public Spectacle on Its Own", in: Rosso, Michela (ed.), *Investigating and Writing Architectural History: Subjects, Methodologies and Frontiers*, Turin 2014, 233-241.

54 As for instance in Bougeois, Henry, "La grande gare du New York Central", in: *La Nature*, 5 May 1906, 353-355, or in Kuentz, L., "La plus grande gare du monde", in: *La Nature*, 21 June 1913, 43-44.

office buildings, presented in the first type of journals, was more of a curiosity and not seen as an example to follow, Grand Central Station, presented in the other type of journals, was immediately analysed according to its value as a future reference to Paris.

As such, the distinctions made in Paris between different elements of New York were extremely clear. Moreover, the all-encompassing notion of New New York was perfectly absent from the multitude of French articles on the American metropolis. Seen from across the Atlantic, New York had not become *the* New New York but merely an increasingly interesting place, an ensemble of disjointed elements forming the imaginary world of Parisians. But this was not really a surprise. After all, it had been New Yorkers who tried hard to build out of diverse and opposing positions a specific and coherent urban identity capable of sustaining the competition with others and ready to place New York's urban culture in a category alongside that of Paris and London. In this local context, the notion of New New York had worked well. Its dichotomy, i.e. contradictory elements of dignified harmony and material struggles, allowed for an inclusive discursive world building process. This in return enabled a city whose material building was disjointed to be defiant and to create the idea of a common whole. Today, this discursive mechanism invented in New York over one hundred years ago still echoes in openly ambitious cities around the world.

LIST OF FIGURES

Fig. 1-2: Wheatly, W. W., "Transporting New York's Millions", in: *The World's Work*, May 1903, 3427, 3430.

Fig. 3: Eaton, Walter Prichard, "Manhattan: An Island Outgrown", in: *American Magazine*, July 1907, 250.

Fig. 4: Cunniff, M. G./Goodrich, Arthur, "The Rebuilding of New York", in: *The World's Work*, December 1901, 1487.

Fig. 5: "What New York Pays To Get To Its Office", in: *The World's Work*, October 1902, 2684.

Fig. 6-9: Blackshaw, Randall, "The New New York", in: *The Century Magazine*, August 1902, 492, 495, 497, 498.

Fig. 10: Forbes, Edgar Allen, "The Skyscraper", in: *The World's Work*, May 1911, 14387.

Fig. 11-14: Baxter, Sylvester, "The New New York", in: *The Outlook*, 23 June 1906, 418, 421-423.

Fig. 15-16: Thompson, Hugh, "The Greatest Railroad Terminal in the World", in: *Munsey's Magazine*, April 1911, 29, 32.

Fig. 17-19: Bruère, Robert W., "The Gates of New York", in: *The Outlook*, 27 April 1907, 928, 931, 929.

BECOMING A PERMANENT CONSTRUCTION SITE. ISTANBUL AFTER 1923

Ipek Şen, Şebnem Şoher

1. INTRODUCTION

With the proclamation of the Turkish Republic in 1923, and after serving as the capital of Ottoman Empire for 500 years, the title passed from Istanbul to Ankara. Though Istanbul lost its position and role in the social, cultural and economic structure of the country, its public image remained virtually unchanged. Situated in the west of the country, and therefore perceived as a gateway to the "West", Istanbul is the centre of trade and accommodates the Palace and the court. The city had always been exceptional when compared to the rest of Turkey, particularly in the areas of education, economy, cultural life and diversity of population. In Turkish literature Istanbul is always held as an example of positive urban culture. The city is regarded as a city of consumption,[1] a cosmopolis and a symbol for "the new" and "the modern".

Our study has departed from the argument that Istanbul has for a long time been the actor of the grand narratives of government due to the above-mentioned characteristics and its public perception. Remarkably, looking at the urban history of Istanbul, these "grand narratives" come with several spatial interventions that claim either to contribute to the city's historical meaning or to change it fundamentally. In this paper, in order to illustrate this argument, we have chosen two consecutive periods that featured big spatial redevelopments of Istanbul's city centre that can also be seen as the starting points for a series of reconstruction projects witnessed in Istanbul right up until today. While these two periods have significantly different discourses in terms of politics, they both perform spatial acts emerging from similar approaches that aim to transform the city and its public image through a series of grand reconstruction projects.

The first part of the study, known as the Prost period in the Turkish literature of architecture and urban planning,[2] focuses on the era that starts with the proclamation of the Turkish Republic and the relocation of the capital in 1923 and ends with the elections in 1950. The prominent project of this era was the Prost Plan, which was prepared by the French architect and urban planner Henri Prost, invited by the government to reconstruct the city centre according to the principles of a modern city. This included an entirely new zoning system consisting of parks and promenades that

---

1 Kemal, Yahya, *Aziz İstanbul, 1000 Temel Eser*, Istanbul 1969, 174.

2 Akpınar, İpek, "The Making Of a Modern Pay-i Taht In Istanbul: Menderes' Executions After Prost's Plan", in: Bilsel, Cana/Pierre, Pinon (ed.), *From The Imperial Capital to the Republican Modern City: Henri Prost's Planning of Istanbul (1936-1951)*, Istanbul 2010, 167-200. 167.

traversed the city centre, wide boulevards which crossed the old city and projected an image that reflected the aims of a nascent Turkish Republic.

The second part of the study focuses on the period beginning right after the first multi-party general and municipal elections of 1950 that resulted in a Democratic Party win and ending with the Military Coup in 1960. This period is named after the Prime Minister Adnan Menderes and the reconstruction projects conducted after 1956 were referred to as "Menderes Reconstructions"[3].

The names of the periods are remarkable in terms of history of urban planning as the first period was named after an urban planner and the second period was named after a political actor. In terms of urban planning practice, it is usually the planner, architect or the mayor of the city who is considered to be the main actor. Yet here, the prime minister filled the role of the planner/architect/mayor by being both the leader and the visible face of the reconstruction projects. Moreover, he was later given the title of "Honorary Mayor of Istanbul"[4]. The point that we try to make here is not that the projects lacked planning and participation, which are equally important, but that they indicate the main discursive difference between these two consecutive eras in Turkish socio-political history.

What is worthy of attention is that the two selected periods present a series of characteristics that are both similar and different at the same time. First, both sets of reconstruction projects targeted Istanbul, the city with the greatest symbolic and economic value. Second, they both pursued an image of a modern city but they differed significantly when it came to defining it. Third, the projects possessed a similar approach of physical planning and social engineering.[5] That is, the aim of both of the projects was to transform the social structure and everyday life of the city by transforming its urban space.

David Harvey, a renowned geographer and anthropologist, refers to this approach as an "imagination" problem. In his book *Social Justice and The City* he presents two different approaches of urban phenomena: sociological imagination and geographical imagination. He argues that choosing only one of these imaginations to understand the city and its society would not be enough to comprehend and solve its problems:

> There are, therefore, signs of some pressure for bringing the sociological and geographical imaginations together in the context of the city. But it has been a struggle. More often than nor geographical and sociological approaches have been regarded as unrelated or, at best, as viable alternatives to the analysis of city problems. Some have sought, for example, to modify the spatial form of a city and thereby to mould the social process (this has typically been the approach of physical planners from Howard on). Others have sought to place institutional constraints on social processes in the hope that they alone will be enough to achieve necessary social goals.[6]

3 Boysan, Burak, "İstanbul'un Sıçrama Noktası", in: Akpınar, İpek (ed.) *Osmanlı Başkentinden Küreselleşen İstanbul'a: Mimarlık ve Kent, 1910-2010*, Istanbul 2010, 81-96, 82.

4 Tekeli, İlhan, *İstanbul'un Planlanmasının ve Gelişmesinin Öyküsü, Tarih Vakfı Yurt Yayınları*, İstanbul 2013, 176.

5 Tekeli, İlhan, *Modernite Aşılırken Kent Planlaması*, Istanbul, 2001, 10.

6 Harvey, David, *Social Justice and the City*, Oxford, 1988, 26.

Similarly, John Friedmann in *Planning in the Public Domain*[7] also mentions planning as a way of social engineering. The Prost Plan as well as the Menderes Reconstructions are both good examples of social engineering. However, neither of these spatial interventions were completed. When the Democratic Party came to power, the implementation of the Prost Plan ceased, the plan was revised by an entirely new group of experts and then eventually shelved. In 1960, due to budgetary problems and the Military Coup, the Menderes Reconstructions shared the same fate as the Prost Plan and could not be finished.

In the Prost Plan, the European-style modern nation-state envisaged by the Republican People's Party was evident through Prost's drawings, influenced by Haussmann's reconstruction of Paris. The aim of the plan was to create public spaces that would redefine the Ottoman city and give it a new European face. However, the Menderes Reconstructions, while based on revisions of the Prost Plan, followed another socio-spatial agenda. This time, the image of people walking along promenades was replaced by wide American-type boulevards for fast automobiles. The Prost Plan proposed a pedestrian-friendly city, developing inside of the city walls and not exceeding the limit of a population of three-and-a-half million. Prost, however, demolished around 1000 buildings to open up large boulevards and public spaces on the Historic Peninsula and was therefore also criticised by the intellectuals of the country. The Menderes Reconstructions, finding Prost's plan insufficient and faulty in many aspects, ripped open the city walls to accommodate 50-metre-wide motorways, demolished 7300 buildings, replaced the parks with high-rise buildings, pulled the city towards its outskirts and triggered a rapid increase in the built environment that eventually resulted in sprawl.

In short, these two periods, two discourses and two big spatial interventions selected for scrutiny by the study are equally important to fully comprehend Istanbul's current situation and problems. From our point of view it is noteworthy that these two projects, though possessing similar approaches and the dream of a modern city, produced quite different outcomes in terms of urban form and urban history. Our aim is to show how both projects emerged from political discourses of the elite that failed to evaluate currents inside society and tended to overlook the needs and the capabilities of the city. The result was therefore a break in urban form and inconsistencies in the collective memory of the inhabitants.

To thoroughly illustrate this argument, this paper will try to shed light on the socio-political atmosphere of the periods in which these reconstruction projects took place and try to compare both the events and the outcomes by positioning Istanbul's urban form at its centre.

2. A NEW CITY FOR A NEW COUNTRY: EARLY YEARS OF THE REPUBLIC AND THE PROST PLAN

The nineteenth century was a period of transformation for the imperial capital of Istanbul. A growing population and severe damage caused by the big

7 Friedmann, John, *Planning in the Public Domain. From Knowledge to Action*, Princeton 1987, 58.

fire[8] required construction of new neighbourhoods. The physical characteristics of the city were transformed for many different reasons: on the one hand, the need for buildings that would resist fire and roads fit for public transportation vehicles rather than horse-carriages; on the other hand, a new spatialisation with an early segregation of social classes, new public spaces and, most importantly, the desire to possess a number of innovations that would symbolize technical and cultural development. As a result, new construction technologies were adopted and many infrastructural investments were made.[9] The fire of 1870 in the Pera district was followed by the emergence of new "western-oriented" recreational areas, such as Tepebaşı Park.[10] The effects of western influence on the transformation of the Ottoman Empire could be seen both in the institutions of the government and the redevelopment of the physical environment.[11] The first scaled maps of Istanbul date from the beginning of the nineteenth century. During the same period, development plans of the city were made via the contribution of foreign experts, such as Melling and Von Moltke.[12] The depth of their influence on the realizations is questionable, but their presence indicates an enthusiasm for western models, previously shown by the Ottoman bureaucrats who equated "Western concepts and institutions [...] with progress and modernization"[13].

The first decades of the twentieth century included several attempts to develop the institutional structure and infrastructural systems of Istanbul through the introduction of new administrative institutions and technologies by dismissing Islamic traditions and adopting European models instead. Detailed maps of the city including cadastral data and physical conditions were drawn with the support of western corporations.[14] In particular, Goad Insurance Maps (1904-1906), German Blue Maps (1913-1914) and, later, Pervititch Insurance Maps (1922-1945) all show the geographical shift of housing areas in the city. Relocation of the government and its administrative components to the Beşiktaş district on the European shores of the Bosphorus and the abandonment of the historic peninsula, followed by the development of new neighbourhoods both on the European side and along the Marmara Sea coast, were indicators of a new socio-economic and urban organization that not only changed the historic city forever but, above all, its perception/image/representational value. From the mid-nineteenth century onwards, the locus of political power repeatedly shifted backwards and forwards from the Pera district to the historic peninsula.

---

8 Çelik, Zeynep, *The Remaking of Istanbul. Portrait of an Ottoman City in the Nineteenth Century*, Berkeley 1993m 52-53. Zeynep Çelik states that 1633 was the date when a series of serious fires started in Istanbul. During the years from 1633 to 1839, 109 fires occurred and from 1853 to 1906 the number of total fires reached to 229.

9 Tekeli, *İstanbul'un Planlanması* (fn. 4), 41. Akın, Günkut, "20. Yüzyıl Başında İstanbul. Toplumsal ve Mekânsal Farklılaşma", in: Akpınar, İpek (ed.), *Osmanlı Başkentinden Küreselle en İstanbul'a: Mimarlık ve Kent, 1910-2010*, Istanbul 2010, 20.

10 Marmaras, Emmanuel V./Tsilenis, Savvas, "Parallel Routes. Proposals For Large Scale Projects in The Centres of Athens and Istanbul at The Beginning of the Twentieth Century", in: *ITU A|Z Journal*, vol. 8, 2011-1, 68-84.

11 Çelik, *The Remaking of Istanbul* (fn. 8), 37.

12 Tekeli, *İstanbul'un Planlanması* (fn. 4), 65; Çelik, *The Remaking of Istanbul* (fn. 4), 50.

13 Çelik, *The Remaking of Istanbul* (fn. 8), 37.

14 Akın, "20. Yüzyıl Başında İstanbul" (fn. 9) , 21.

At the beginning of twentieth century, when Joseph Antoine Bouvard was at the peak of his career and in charge of the Architectural Department of Paris, Abdülhamit II was very impressed by his exposition design and commissioned Bouvard to design a city plan. Unsurprisingly, his partial proposal concentrated on just a few areas in the historical city in order to create a new and modernized image.[15] Later, in 1910, André Auric was invited from Lyon to develop infrastructure in Istanbul. Again, his ideas, inspired by the principles of The City Beautiful Movement, were criticised for not being specific enough and only covering some parts of the city.[16]

In the years following World War I, when the Turkish Republic was formed, Ankara became the new capital of the nation and the government attempted to create a Turkish identity by denouncing the Ottoman legacy. This political decision led to the official neglect of Istanbul, since most public finance was being invested in the new national capital.[17] In this period, the Turkish architectural scene was dominated by the discussion of breaking with Ottoman revivalism and embracing the modern movement, called "new architecture"[18]. In order to construct the new organizational structure of the state and its modern identity, German professionals in particular were invited, consulted and commissioned after 1924, the date when a Treaty of Friendship was signed between Turkey and Germany.[19] Ankara witnessed the construction of many public buildings and the introduction of new urban legislations during the 1920s and 1930s. Finally, in 1930, as architectural and urban historian Murat Gül mentions, a new municipal law was invoked that entailed the foundation of new city administrations and demanded city plans for all cities with more than 10000 inhabitants.[20] Although the names of Emin Erkul (the mayor of Istanbul between 1924-1928) and Carl Lörcher (who developed an early urban planning scheme for Istanbul) are mentioned in relation to urban interventions,[21] the delayed attempts at acquiring a general plan for Istanbul became a pressing matter by the 1930s. In 1932, four planners, Donald Alfred Agache, Hermann Ehlgötz, Jacques Henri Lambert and Henri Prost, were invited to participate in a competition. Ehlgötz won, but his plan was never realised. Following a long tradition of inviting western experts, Martin Wagner was commissioned as planner and consultant. Finally, in 1936, Henri Prost, though not among the winners of the competition, was commissioned to prepare a plan for the city[22] as a result of his experiences in Casablanca, Marrakesh and Rabat.[23]

15 Çelik, *The Remaking of Istanbul* (fn. 8), 111.

16 Gül, Murat, *The Emergence of Modern Istanbul. Transformation and Modernisation of a City*, London 2009, 67.

17 Bilsel, Cana, "Remodelling the Imperial Capital in the Early Republican Era: the Representation of History in Henri Prost's Planning of Istanbul", in: *Politica* 306, 2007, 98.

18 Bozdoğan, Sibel, *Modernism and Nation Building. Turkish Architectural Culture in the Early Republic*, Seattle 2001, 21.

19 Akcan, Esra, *Çeviride Modern Olan. Şehir ve Konutta Türk-Alman ilişkileri*, İstanbul 2009, 35.

20 Gül, *The Emergence of Modern Istanbul* (fn. 16), 80.

21 Tekeli, *İstanbul'un Planlanması* (fn. 4), 122.

22 Ibid., 135.

23 Boysan, Burak, "Aron Angel ile İstanbul. Prost ve Planları Üzerine Söyleşi", in: *Mimarlık*, vol. 285, 1999, 30-40, 30.

Henri Prost proposed a plan to be realised in two stages. The execution of the first phase began in 1939 and continued until 1948; the second phase covered the period from 1943 until 1953. Cana Bilsel interprets Prost's work in Istanbul as a plan for a general transformation of the city, on the grounds of his lectures held in Paris in 1947.[24] Prost prioritised sufficient public transportation infrastructure for the expanding outskirts, hygienic housing environments and public spaces.[25] There were two significant motorways proposed in the plan. The first one started from Taksim Square, crossed Golden Horn and the whole historic peninsula and ended at Yenikapı. The second road, using Galata Bridge, connected Taksim Square with Beyazıt Square, thereby crossing Istanbul's former business district of Eminönü.[26] A new coastal highway and the conversion of the suburban railway into underground transportation were also parts of the plan. The relocation of industrial facilities to the shores of Golden Horn and Marmara Sea, beyond the city walls, contributed to the rehabilitation of housing.[27] Important highlights of the plan were, an Archaeological Park(covering an area near Sarayburnu and Sultanahmet, including Topkapı Palace and Byzantine Hippodrome), Park No: 1 (located outside of the city walls near Topkapı, and would embrace former vegetable gardens and mixing them with educational functions such as botanical garden and zoo) and finally Park No: 2 (extending from Taksim Square to Dolmabahçe through Maçka district, combining green area with cultural and sports facilities, such as open-air theatre, stadium, exhibition hall).[28] In addition to these major interventions, the rehabilitation and redesign of Taksim, Beyazıt and Üsküdar Squares, public beaches and new boulevards were all components of the Prost Plan.

3. MENDERES PERIOD (1950-1960): THE GOLDEN YEARS OF MOTORWAYS AND URBAN SPRAWL

To understand the reconstruction of Istanbul in the second half of 1950s, which was led by Prime Minister Adnan Menderes, a closer look at the background of the country is necessary. As during the 1950s in most developing countries, the reconstruction of historic cities was in fashion.[29] Menderes' gigantic project, targeted at the greatest city in Turkey, also followed this trend.

As Hobsbawm terms it, in his book *The Age of Extremes*, the 30 years between 1950 and 1980 are commonly described as the Golden Years in world history, imbued with promises of a more democratic and open culture, the end of starvation, an increasing population and the abundance of food

24 Bilsel, Cana, "'Les Transformations d'Istanbul.' Henri Prost's planning of Istanbul (1936-1951)", in: *ITU A|Z*, vol. 8, 2011, 100-116, 100.

25 Bilsel, Cana, "İstanbul'un Dönüşümleri. Prost Planlaması ve Modern Kenti Yaratmak", in: Akpınar, İpek (ed.), *Osmanlı Başkentinden Küreselleşen İstanbul'a. Mimarlık ve Kent, 1910-2010*, Istanbul 2010, 49-67, 53-7.

26 Bilsel, *Remodelling the Imperial Capital* (fn. 17), 104.

27 Bilsel, Cana/Zelef, Haluk, "Mega Events in Istanbul from Henri Prost's Master Plan of 1937 to the twenty-first-century Olympic Bids", in: *Planning Perspectives*, vol. 26, 2011, 621-634, 626.

28 Bilsel, "Les Transformations d'Istanbul" (fn. 17), 107.

29 Akpınar, "The Making" (fn. 2), 167. Akpınar states that in 1950's in eastern Mediterranean Countries similar urban reconstruction projects were started: Konstantinos Karamanlis in Greece, Gamal Abdel Nasser in Egypt, David Ben-Gurion in Israel and Riza Shah and his son Muhammed Riza Shah in Iran.

and other goods.[30] For the young Republic of Turkey, already tired of the ramifications of two world wars and the conservative politics of a one-party government, these years were thought to herald a new era.

The year 1950 was a real watershed in the history of the Republic of Turkey. It was the year of the first multi-party elections and a number of serious democratic changes in constitution.[31] In 1950, the Republican People's Party (CHP), after holding power for 22 years, lost the elections, which resulted in the Democratic Party (DP) forming a government, a party which had been in opposition since its foundation in 1946. Following the elections, Adnan Menderes, whose name was later given to the Istanbul reconstruction projects, became Prime Minister.[32]

For the Republic of Turkey, it was a time of change. The conservative politics of CHP were no longer meeting the needs of the public and the new era. Industrialisation and urbanisation accelerated, while new social classes with new demands emerged. Therefore, when the DP came to power, the first period of their government (1950–1954) as a result of both the liberal economic model the party followed and their open and inclusive politics and the public enjoyed rapid improvements in agriculture, industry, urbanisation and a significant rise in purchasing power.[33]

The DP was focused on growth. The government allocated most of its resources to economic development and envisioned progress with big reconstruction projects, agricultural exports and industrialisation.[34] The remarkable economic growth between 1950 and 1954 (13 per cent)[35] led to another success in the second elections. The party won with a higher percentage of the votes than in the first elections.[36] However, this success came with consequences. The liberal economic model caused a dependence on foreign sources and the end of the Korean War resulted in a decline in grain exports.[37] Furthermore, a severe drought caught the country off guard and last-minute efforts to avoid a decline in growth failed: production rates dropped and Turkey headed into recession.[38] The population working in agriculture dropped from 84,1 per cent to 76,8 per cent in five years (1950–1955)[39] and this decline resulted in migration to industrial zones, Istanbul being one of the main destinations.

Istanbul was the biggest city in Turkey at this point, with great symbolic value that referred back to the years of Ottoman Empire. It also had significant economic value with job opportunities generated by new

30 Hobsbawm, Eric, *The Age of Extremes*, London 1995, 260.

31 Makal, Ahmet, *Türkiye'de Çok Partili Dönemde Çalışma İlişkileri 1946-1963*, Ankara 2008, 6

32 Ertem, Barış/Altunel, Mustafa Cevdet, "İstanbul İmarındaki Tarihi Eser Kaybının Tarih ve Turizm Açısından İncelenmesi. Karaköy-Kabataş Bölgesi", in: *Sosyal Bilimler Araştırmaları*, vol. 2, 2011, 61-79, 65 (as cited in İhsan Ezherli, TBMM: 1920-1986, Ankara 1986, 94.)

33 Makal, *Çalışma İlişkileri* (fn. 31), 8; Ertem/Altunel, "İstanbul'un İmarı" (fn. 32), 63.

34 Baytal, Yaşar, "Demokrat Parti Ekonomi Politikaları (1950-1957)", in: *Ankara Üniversitesi Türk İnkılap Tarihi Enstitüsü Atatürk Yolu Dergisi*, vol. 40, 2007, 551 (as cited in TBMM Zabıt Ceddesi, D: 10, c: I, 14.5.1954, 26).

35 Ertem/Altunel, "İstanbul'un İmarı" (fn. 32), 65.

36 Ibid., 66.

37 Baytal, "Ekonomi Politikaları" (fn. 34), 556.

38 Özçelik, Pınar Kaya, "Demokrat Parti'nin Demokrasi Söylemi", in: *Ankara Üniversitesi SBF Dergisi*, vol. 65, 2010, 180.

39 Baytal, "Ekonomi Politikaları" (fn. 34), 558 (as cited in 1954 Bütçe Kanunu Layihası, Ankara, 1954, 3-64).

industrial investments. Migrating populations surged towards the city, though, and Istanbul was not prepared. Prime Minister Menderes criticised the former government for its neglect of Istanbul. For him, a comprehensive programme to redevelop Istanbul was a long-overdue necessity. Therefore the 500th anniversary of the conquest of Istanbul served as a well-timed trigger to start reconstruction.[40]

4. THE REDEVELOPMENT OF ISTANBUL

Istanbul's reconstruction is usually regarded as having occurred during two separate periods. The first starts with the decommissioning of Henri Prost and revisions of the Prost Plan, which continued until right after the second elections; the second addresses the period between 1956 and 1959, the years when economic difficulties started to become visible.[41]

4.1 FIRST PERIOD: 1950-1956

Henri Prost was decommissioned on 26 December 1950, but his plan was not shelved immediately. The government disapproved of the idea of inviting foreign experts to solve the problems of Turkish cities.[42] This approach of the former government also contradicted the emphasis placed by the new government on "Turkishness"[43] so a new group comprising Turkish planners, architects and consultants was employed to revise Prost's plan. The revision report found many faults in Prost's work: the topographic maps were inadequate, the projections for housing, transportation, economic and social structure were poorly developed and instead of finding solutions for a rapidly developing metropolis, the plan was accused of being focused only on the beautification of the city.[44] Prost was known for his fondness of Hausmann's Paris and in his plan he proposed 18 parks, many entertainment areas, an Olympic zone, university buildings and the widening of former squares to create public spaces.[45] Although he also suggested boulevards of 50 metres wide to connect the inner city to its outskirts and construct new motorways and bridges to create a metropolis from a former Ottoman pedestrian city, these aspects of his plan remained in the background.

The Menderes period was usually criticised for its lack of plans but the first period of reconstruction can be described as a period of too many plans. The re-visioning committee tried to draw up several plans to regulate industrial and commercial zones, new housing projects and harbours, but immigration from Anatolia and the rapid development of industry outpaced

40 Akpınar, "The Making", (fn. 2), 172

41 Özçelik, Pınar Kaya, "Demokrat Parti'nin Demokrasi Söylemi", in: *Ankara Üniversitesi SBF Dergisi*, vol. 65, 2010, 163-187, 180.

42 In spite of this approach, the government invited Hans Högg to revise Istanbul's plans in 1956. Tekeli, *İstanbul'un Planlanmasının* (fn. 4), 179.

43 Akpınar, "The Making" (fn. 2), 171.

44 Tekeli, *İstanbul'un Planlanmasının* (fn. 4), 171.

45 Akpınar, İpek, "Sunuş", in: Akpınar, İpek (ed.), *Osmanlı Başkentinden Küreselleşen İstanbul'a: Mimarlık ve Kent, 1910-2010*, Istanbul, 2010, 11.

the planning process[46] and redevelopments in this period were made in moderation.[47] Decisions about the placement of the harbours, connections with the Anatolian commercial roads and the location of industrial facilities were made carefully in order to create a functioning industrial city.

4.2 SECOND PERIOD: 1956-1961

In 1956, both the pace and scope of reconstruction changed significantly. The economy of the country was no longer in as good a shape as it had been in the early 1950s. The public started to show discontent and the country faced difficulties in finding new external loans.[48] Furthermore, the 1957 elections were imminent and the city of Istanbul now had one million inhabitants, representing an expanded electorate.[49]

Regardless of these problems, the government continued with its programme because the promise of progress required solid developments that would satisfy the public and prevent losing support at the elections Therefore the heavy investment continued. At a press conference on 23 September 1956, Prime Minister Menderes announced the Istanbul Programme in the following words:

> [...] We could not get the 300 millions from Americans. The two yearlong drought is very likely to last for another year. And we will start to import grain. However, all the indigenous resources of the country are actuated. In the end Turkey will be a self-sufficient country. We will realise these projects even though it is said that the state's and the government's power and resources were depleted.[50]

In 1956, as the Prime Minister himself described at the press conference, the 300 million dollar credit request by the Turkish Government from the US was declined; the country was in a difficult position and in debt because of the drought and the investments made during the course of the DP's first period in office. However, Menderes was determined to show that the country still possessed the power to continue, and refused to cease the redevelopment plan deciding instead to use it as an indicator of progress. The press conference was the departure point of the second half of Istanbul's redevelopment. This time there was no plan or scientific research accompanying the Prime Minister's programme. Tekeli interprets Menderes' approach at this time as "a program shaped by intuitions and possibilities."[51]

The gigantic reconstruction projects in Istanbul were advertised as a re-conquest, a rediscovery of the beautiful city. As Prime Minister Menderes

46 Tekeli, *İstanbul'un Planlanmasının* (fn. 4), 172-175.

47 Boysan, "İstanbul'un Sıçrama Noktası" (fn. 3), 82.

48 Tekeli, *İstanbul'un Planlanmasının*, (fn. 4), 176 and Özçelik, "Demokrat Parti" (fn. 41), 178.

49 Boysan, Aydın, "Adnan Menderes Belediyeciliği İmar Hareketi Uygulama ve Sonuçları", in: *Mimarist*, vol. 3, 2004, 25-31, 25.

50 Tekeli, *İstanbul'un Planlanmasının* (fn. 4), 176 (as cited in "Sayın Başvekil Basına Geniş İzahat Verdi ve İstanbul'un İmar ve Kalkınma Prensiplerini İzah Etti", in: *İller ve Belediyeler Dergisi*, no: 132, 1956, 645).

51 Ibid., 177.

declared at the press conference, the government used all available resources to achieve this and even created new ones. Approximately 536 million Turkish Lira were allocated during the reconstruction of the city between 1956 and 1960.[52]

The Prost Plan was criticised by the plan revision committee for its lack of comprehensiveness and its focus solely on beautification. However, Istanbul's next urban renewal project followed a similar approach. Menderes also held the opinion that no great city could ever be complete without grand squares and statues of the great heroes of the country's past.[53] What he opposed was that the previous government only concentrated on the Beyoğlu area.[54] Therefore he put the old city at the centre of the reconstruction plans. His main concern was the traffic. For Menderes and his consultants, open highways were the indicator of a well-functioning city.[55] Most of the constructions and demolitions were made to open new roads either to connect different parts of the city or to emphasise Ottoman architectural heritage. To open these new roads or widen existing streets, surrounding buildings were demolished, partially torn down or replaced.[56] Even though a great deal of importance is attributed to the architectural beauties of the former empire, in one Aksaray district alone 29 Ottoman-era buildings including mosques, fountains, madrasas and hammams were demolished [fig. 1]. This number does not include other demolished buildings such as houses, synagogues and churches.[57]

Fig. 1 Aksaray, Vatan and Millet Roads.

52 Ertem/Altunel, "İstanbul'un İmarı" (fn. 32), 72 (as cited in İhan Tekeli-İlber Ortaylı (1978). Türkiye'de Belediyeciliğin Evrimi, Ankara: Türk idareciler Derneği, 174-175).

53 Tekeli, *İstanbul'un Planlanmasının* (fn. 4), 177.

54 Boysan, "Sıçrama Noktası" (fn. 3), 93.

55 Ibid., 88.

56 Ibid., 83 and Ertem/Altunel, "İstanbul'un İmarı" (fn. 32), 72.

57 Boysan, "Sıçrama Noktası" (fn. 3), 84.

A great number of roads opened during the second half of the project but there are several important ones that dramatically transformed the urban form of the historical Istanbul. The first (I) one was the Sirkeci-Florya Costal Road that surrounded the old peninsula where the Topkapı Palace was located. Prost was also planning a coastal highway in this area but his intention was to build the road inside the walls to accentuate the silhouette of the city walls and not to block the city's access to water.[58] The second (II) one was London Asphalt [fig. 2], that was to go to the fringes around the Küçükçekmece Lake where, in the following years, new housing projects rapidly developed towards the westwards. The third (III) is Millet, 50 metres wide and connected to the London Asphalt and its parallel Vatan Boulevard that replaced the Prost's Park No: 1 and Botanical Garden.[59] [fig. 1] On the European side of the Bosporus Tophane-Karaköy-Azapkapı Road (IV) was widened by demolishing thousands of houses and a considerable number of historical buildings [fig. 3] and another 50-metre wide road (V), Barbaros Boulevard, opened towards the northern part of the city that caused the sprawl to the north. Over time this sprawl created today's CBD Maslak and stinye, where at that time no settlements were located and even the district had no name.[60]

Until the 1958 devaluation, when the Turkish Lira lost 325 per cent of its value, the reconstruction projects were carried out simultaneously and treated as monumental events that would transform Istanbul into a modern

Fig. 2 Edirnekapı.

58 Boysan, "Aron Angel" (fn. 23), 38.

59 Ibid., 38.

60 Tekeli, *İstanbul'un Planlanmasının* (fn. 4), 181-190.

Fig. 3 Tophane, Meclis-i Mebusan Street.

city and Turkey into an advanced country. But after 1958, the projects lost momentum and, in 1959, the resources allocated to urban development only reached 29 per cent of their 1957 value.[61]

In just three years, the Menderes Reconstructions made a great impact on Istanbul's form and its social environment. As reconstruction and demolition continued without a proper plan, immigration from Anatolia gained pace and immigrants settled on empty areas close to the industrial zones and the newly constructed roads. New illegal settlements (gecekondu) followed one another while displaced inhabitants of the historic city centre also looked for new places to live. New high-rise apartments were constructed along the new highways resulting in sprawl and suburbanisation. After 1959, the pending projects were shelved due to economic difficulties leaving the city a huge construction area with more problems than ever before.

Similar to the former government's approach, the transformation of the social structure through the reconstruction of urban space was one of the aims of the government. In this aspect, the writings of Semavi Eyice, one of the members of the re-visioning committee, were notable. He hoped that Istanbul's reconstruction would educate the city's inhabitants: a person walking these wide straight roads, he notes, is protected from dark thoughts, immorality and will approach life with an open mind and modern thinking.[62]

The result of Prime Minister Menderes' approach and the hastily-made decisions was an unnatural and unhealthy urbanisation,[63] uncontrolled

61 Boysan, "Sıçrama Noktası" (fn. 3), 90.

62 Ibid., 85.

63 Sayar, Zeki, "İmar ve Eski Eserler", in: *Arkitekt*, vol. 26, 1957, 49-50, 49.

sprawl, illegal housing, rapid increase in automobiles and traffic problems[64] and consequently an irreversible change in the form of the city. As urbanisation and industrialisation were inevitable, what interrupted the city's natural development was the lack of programming and, as Tekeli puts it, scientific research that supported the projects.[65] The four-year long operation disregarded various existing problems and the potential of the city, causing an imbalanced growth in the city's structure and the desecration of its form.

5. CONCLUSION

Our research covers the Republic of Turkey's early years from the mid 1930s until the end of the 1950s during which period the country went through many changes in a short time: the repercussions of the fall of the Ottoman Empire, the foundation of the Republic, the transition from a one-party regime to a partial democracy and the introduction of a liberal market economy. In its first 25 years, Turkey witnessed some of its most turbulent events.

In those years, modernism, nation-state, liberalism, market economy, freedom and welfare were concepts that had taken hold of the world. Both the government and the population of Turkey were also strongly influenced by this global transition. The wars were over, the economy was picking up and governments entertained the idea that Turkey, with its social and physical appearance, should keep up with these worldwide changes.

The two reconstruction projects we presented in this study were chosen to portray this "need for change". The first project, started in the early years of the Republic of Turkey, characterised symbolic construction of a new regime that emphasised the nation and the nation-state. To redevelop Istanbul according to the new regime's needs, the new national government invited many European architects and planners to become involved. It is worth noting, however, that while the new regime was known for denouncing the legacy of the Ottoman Empire, it, perhaps unconsciously, mimicked the methods of Ottoman bureaucrats who were very keen on seeking help of western experts as the western models of urban planning were being associated with modernisation and progress during the nineteenth century.[66]

The plan that Prost prepared for the transformation of Istanbul was chosen from many other competition entries. It was the most convenient plan for reflecting the notions of the new government with European-like beautification and the renewal of the city by introducing "hygiene, aesthetics and circulation"[67] to the former capital's everyday life.

The Menderes Reconstructions were originally based on the revisions of the Prost Plan but gained a different character over time. DP government, which criticised the Prost Plan for its lack of vision and not being based on thorough-enough investigations or accurate topographic maps, followed the same path and did not propose a new approach to the planning process.

64 Akpınar, "Sunuş", (fn. 45), 11.

65 Tekeli, *İstanbul'un Planlanmasının* (fn. 4), 177.

66 Çelik, *The Remaking of Istanbul* (fn. 8), 37.

67 Bilsel, *İstanbul'un Dönüşümleri* (fn. 25), 57.

The discourses in each project also demonstrated very similar characteristics. Both periods lay emphasis on the transformation of the old city into a new metropolis. Towards the end of the second phase of the Menderes Reconstructions, the city became the site of daily demolitions that were presented as spectacles of progress and transformation with an emphasis on "little America" and "automobiles riding smoothly on newly paved roads"[68]. The consequences of the Menderes Reconstructions were immense: in just 60 years the population of Istanbul multiplied by 12, its area by 30 and the number of automobiles by 900.[69]

Both eras had differences in discourse and implementation, rooted in the distinct political stances of the two parties; however, they produced amazingly similar results for Istanbul's urban form and everyday life. Whether it was the beautification of the physical environment or the opening of wide boulevards for the sake of automobiles, disruption to the natural course of events has always had consequences. In the case of Istanbul, demolishing existing residential areas, historic buildings and monuments from different periods and transforming streets into boulevards or squares resulted in an inevitable transformation of the everyday routines of inhabitants. People who had lived in the city centre for long periods had to move to the outskirts, the meeting points, the references and landmarks of the city changed and the new inhabitants, without proper introduction, were forced to adapt to daily life in this newly-formed city. As a result, memories and meanings attached to urban spaces were wiped from the city's historic and spatial scene.

These consecutive events show us that plans based on ideology and "improvement" rather than the needs of the inhabitants and the city are always prone to cause unwanted results that affect not only the present but also future experiences. Collective memory impairment, deformation of the urban form and transformation of everyday routines of a city produce recurring and persistent issues for urban space and future inhabitants of the city.

LIST OF FIGURES

Fig 1: Deutsches Archäologisches Institut Istanbul, D-DAI-IST-KB7442.

Fig. 2: Deutsches Archäologisches Institut Istanbul, D-DAI-IST-R22781.

Fig. 3: Deutsches Archäologisches Institut Istanbul, D-DAI-IST-R32772.

68 Akpınar, "Sunuş" (fn. 45), 11.

69 Boysan, "Sıçrama Noktası" (fn. 3), 81.

NATIONS UNDER CONSTRUCTION

BUILDING A CATHEDRAL FOR THE NATION

POWER HIERARCHIES, SPATIAL POLITICS AND THE PRACTICE OF MULTI-ETHNICITY IN INTERWAR CLUJ, ROMANIA (1919-1933)[1]

Liliana Iuga

1. CATHEDRALS, NATION AND POWER

Since the Middle Ages, cathedrals have represented important urban landmarks that shaped the space and identity of European cities. During the nineteenth century these places traditionally associated with power and prestige were transformed through romantic nationalism into national symbols.[2] In the words of the famous French architect and pioneer of the restoration movement Eugène Viollet-le-Duc, "cathedrals [in northern France] were religious monuments, but also national edifices, [...] the symbol of French nationality, and the first and most powerful attempt to achieve unity."[3]

Ironically, before becoming such "national edifices" in the Romanian and more specifically Transylvanian context, Orthodox cathedrals needed to be first constructed. In a region whose population was divided along ethnic and religious lines, the numerous Gothic and Baroque churches dominating the landscape of Transylvanian towns reflected specific power hierarchies. The legal order of (late) medieval Transylvania, part of the kingdom of Hungary, was based on the system of three privileged nations (noble Magyars, Szeklers and Saxons) and four religions (Catholic, Lutheran, Calvinist and Unitarian). Urban churches had therefore been constructed mostly by Hungarians and Saxons, whose religious affiliation was either Catholic or Protestant. By contrast, despite their high number, Romanians were denied political representation, and the Orthodox Church, to which they belonged, was unprivileged. The exclusion was based on the criterion of social class, as Romanians were mostly peasants, but also had religious motivations.[4] The end of the First World War marked a significant shift of power and the redrawing of existing borders. Transylvania was incorporated into Romania, which re-emerged as an enlarged state on territories that had belonged to the Russian and Austro-Hungarian Empires.[5] This change of

---

1 This article builds on the research made for my MA thesis defended at the Central European University, Budapest in 2010. I would like to thank my academic advisers, Professor Constantin Iordachi and Markian Prokopovych, as well as my colleague Zsuzsa Sidó, whose comments and support have improved my work in many ways.

2 Vauchez, André, "The Cathedral", in: Nora, Pierre, *Realms of memory*, vol. II, New-York 1997, 56.

3 Quoted in ibid., 58.

4 Hitchins, Keith, *A Nation Discovered: Romanian Intellectuals in Transylvania and the Idea of Nation, 1700-1848*, Bucharest 1999, 12-23.

5 Romania annexed Bukovina, historic Transylvania and territories which were part of Eastern Hungary from the Austro-Hungarian Empire, and Bessarabia from Russia. It was argued that all these territories were inhabited by a Romanian majority. As a result, the Romanian state more than doubled its surface and population. Yet, approximately one third of this population consisted of minorities. Livezeanu, Irina,

sovereignty produced a small, yet significant alteration in the topography of Transylvanian towns and villages: an impressive church-building activity was initiated in the region, with several hundred Orthodox churches being built during the interwar years. In urban areas, these religious edifices took monumental proportions, and came to be known as "cathedrals". [6]

In this paper I will discuss the construction process of the Romanian Orthodox Cathedral in the city of Cluj/ Kolozsvár, in the 1920s and 1930s. The unofficial capital of Transylvania, Cluj had been a stronghold of Hungarian nationalism in the Austro-Hungarian Empire. Unlike other East-Central European cities of the Habsburg Monarchy, such as Prague, Pozsnoy/ Bratislava or Lemberg/ Lviv, nineteenth century Cluj did not face interethnic struggles over the control of public space or local institutions,[7] simply because the Hungarian hegemony was undisputed here, with Romanians occupying a comparatively modest position in the city's social, economic and cultural structures. However, with the territorial changes after the First World War, Romanian state officials felt compelled to remedy this situation through nationalization policies, and increase Romanian presence in the urban space.

To visibly mark that Transylvania belonged to Romania, Orthodox churches were preferred symbols. They were perceived as embodiments of the national identity, as religion and ethnicity were strongly interconnected, with Romanians belonging to the Eastern, and Hungarians and Saxons to Western Christianity. As the American historian Keith Hitchins argues, "in the absence of a privileged order or of representation in the political councils of the land, the Orthodox Church served the Romanians as a 'national' institution."[8] Or, in the words of a group of Transylvanian politicians, "the [Orthodox] Church was the shield that preserved along the centuries our language, traditions and land".[9] Interwar Orthodox churches came to symbolize a past of sacrifice and humiliation, when Romanians had been forbidden to construct such representative buildings within inner-city perimeters, and a present that brought salvation in the framework of a national state that united all the territories inhabited by ethnic Romanians. This case study shows that well into the twentieth century, cathedrals were perceived as visual expressions of a nation- and state-building process, creating a sense of identification for an "imagined political community"[10].

Cultural Politics in Greater Romania, Ithaca 1995, 8-10.

6 An overview of this building activity is provided by Jula, Raluca Diana, *Arhitectura religioasă a românilor din Transilvania în perioada interbelică* ("Religious Architecture of Transylvanian Romanians during the Interwar Period"), Cluj-Napoca 2010. Jula enumerates over 50 Orthodox churches of larger size (67-69).

7 Brubaker, Rogers et. al., *National Politics and Everyday Ethnicity in a Transylvanian Town*, Princeton 2006, 95.

8 Hitchins, *A Nation Discovered* (fn. 4), 18.

9 Arhivele Mitropoliei Clujului (Archives of the Metropolitan Bishopric Cluj-AMC) Fond II-23-920, congratulatory telegram addressed by Transylvanian state officials (Petru Groza, Octavian Goga and Theodor Mihaly) to Nicolae Ivan on the occasion of his birthday. On Orthodoxy in the interwar intellectual debates, see Hitchins, Keith, "Orthodoxism. Polemics Over Ethnicity and Religion in Interwar Romania", in: Banac, Ivo/Verdery, Katherine (ed.), *National character and national ideology in the interwar Eastern Europe*, New Haven 1995, 135-156.

10 Anderson, Benedict, *Imagined Communities. Reflections on the Origin and Spread of Nationalism*, London et. al. 1991, 6.

The existing literature on the construction of Orthodox churches in interwar Transylvania stresses the state's role in supporting such projects, as they were perceived as a means for visually "conquering" the territory through the symbolic significance of the Romanian national style.[11] However, with my research I aimed to highlight the importance of local actors in completing such ambitious construction projects. More than mere representatives of the Romanian state, actors such as Orthodox Bishop Nicolae Ivan in Cluj brought forward their own agenda of increasing Romanians' visibility in the urban topography and "compete" on equal terms with the other ethnic groups and religious confessions. Although the political context was significantly different, I argue that there are important elements of continuity with the pre-war agenda of Transylvanian Romanian elites, as this nationalist project was well grounded into the realities of multi-cultural environment.

2. NATIONALIZATION POLICIES AND EVERYDAY LIFE IN INTERWAR CLUJ

In the interwar period, Cluj had approximately 100,000 inhabitants, about half of whom were ethnic Hungarians, one third Romanians, 13% Hungarian-speaking Jews, and a couple of thousands Germans.[12] Given the modest impact of the Romanian element in urban affairs, state authorities felt that policies of nationalization were a matter of necessity. In the 1920s, the Liberals in power passed a number of laws aimed at creating a centralizing and unifying framework for the new state. As Irina Livezeanu pointed out, similarly to other Eastern European countries in the interwar period, the strategy of nationalization promoted by the Romanian state focused on politics rather than economics, and thus proved more interventionist in the cultural and educational fields.[13] In Cluj, Romanians were named in key positions in administration and culture, and some of the most important Hungarian national institutions, such as the Theatre and the University, were seized by the new state authorities. Such measures were naturally received with great dissatisfaction by the local Hungarians, who felt that their rights and cultural traditions were being strongly disregarded.[14]

Yet, the existing structures proved more enduring than Romanian nationalists had expected. Apart from official contexts, changes in everyday life have occurred more gradually. Dudley Heathcote, an Englishman who visited Cluj in 1925 for example observed that local elites still were mostly

11 Popescu, Carmen, *Le style national roumain. Construire une nation a travers l'architecture. 1881-1945*, Rennes 2004, 211; Ioan, Augustin, *Power, Play and National Identity*, Bucharest 1999, 23-24.

12 http://varga.adatbank.transindex.ro/?pg=3&action=etnik&id=5290 (accessed 6-4-14).

13 Irina Livezeanu's book on the impact of centralization policies in the multi-ethnic provinces of Greater Romania (Bukovina, Bessarabia and Transylvania) points out to the importance of cities, where ethnic Romanians were usually a minority. As such, the state's cultural policies should have focused on the "nationalization of towns, urban elites and cultural institutions". See Livezeanu, *Cultural Politics* (fn. 5), 10, 16-18, 23.

14 Romanian authorities claimed that these institutions were indispensable for the education of the Romanian population. See Ceuca, Justin, *Zaharia Bârsan*, Cluj-Napoca 1978, 103; Livezeanu, *Cultural Politics* (fn. 5), 218-227. For the Hungarian reaction (although from a heavily biased perspective), see Biró, Sándor, *The Nationalities Problem in Transylvania, 1867-1940*, New York 1992, 579-580, 601-603.

Hungarian, „with a sprinkling of Romanians and Germans". However, a strange mixture of Romanian and Hungarian characterized the city for him, from the dishes in the restaurants' menus to the ethnically mixed audiences attending opera performances. Heathcote's account suggests that despite the policies of nationalization carried on from above by the representatives of the Romanian state, the city's social and economic hierarchies remained largely the same.[15] Romanian nationalists agreed with the observation but condemned the situation. Although dominating public institutions, their impact on the city's economic life remained modest in the interwar years.[16] A typical stereotype contrasted the rich non-Romanian merchant to the poor Romanian intellectual.[17] Not least, the public sphere was equally dominated by the Hungarian language press read by everyone, including many Romanians, which naturally enraged the nationalists.[18]

3. NATIONALISM AND SPATIAL POLITICS IN CLUJ

Even before the construction of the Orthodox Cathedral, the design of public space in Cluj was dominated by nationalist agendas and claims for historical continuity legitimizing "the right to the city". [19] In the late nineteenth century, the Hungarian government sponsored the transformation of the main city square from a commercial space into a representative area in the context of the Millennium celebrations, which marked one thousand years of Hungarian presence in the Carpathian Basin. The small shops and houses surrounding the fifteenth century Saint Michael's Church were demolished, making space for a monument representing the Cluj-born King Matthias Corvinus that was unveiled during an impressive national celebration in 1902.[20] The roles were reversed in the interwar period. In order to justify their „right to the city", Romanian nationalists pointed out the city's foundation during Roman times, one thousand years before the arrival of

15 Heathcote, Dudley, *My Wanderings in the Balkans*, London, 1925, 88, 92. Heathcote's account suggests that a compromise solution was adopted even in the case of the nationalized Theatre, with the Romanian Opera House periodically organizing "Hungarian nights" attended by mixed audiences.

16 Buzea, Octavian, *Clujul: 1919-1939*, Cluj, 1939, 219.

17 *Clujul*, 10 August 1924.

18 *Clujul*, 6 January 1924; *Renaşterea*, 17 February 1924. The necessity of "Romanianizing Romanians" as a first step in the nationalization of the city was expressed in an article from 1926 authored by Corneliu Codarcea. The journalist was highly critical towards the attitude of Hungarian-assimilated Romanians who "close themselves in the caves of dark coffee houses, play pool or cards, speaking a slang that is half Romanian, half Hungarian." Despite their ethnic background, stated Codarcea, they refused to integrate into the Romanian public sphere: they did not read Romanian press, avoided people from the Old Kingdom and did not attend Romanian language performances at the Theatre. Corneliu Codarcea, "Kolozsvár-Cluj: Problema românizării oraşelor din Ardeal" ("Kolozsvár-Cluj. The Question of Romanianization of Transylvanian Cities"), *Ţara Noastră*, 27 June 1926, 736-738.

19 The construction of the Orthodox Cathedral in Cluj can be seen as a chapter in the history of spatial politics in Cluj, a broader topic analyzed from a long-time perspective by Margit Feischmidt and Rogers Brubaker: Feischmidt, Margit, *Ethnizität als Konstruktion und Erfahrung. Symbolstreit und Alltagskultur im siebenbürgischen Cluj*, Münster 2003; Brubaker, *Nationalist politics* (fn. 7). Perhaps the best known book on Cluj, the work coordinated by Brubaker focuses on the spatial policies promoted in the 1990s by the ultra-nationalist mayor of Cluj, Gheorghe Funar. The analysis of present-day politics of nationalization is well-grounded into local history. See Part I, Nationalist Politics: Past and Present, 23-166.

20 The nationalist rhetoric constructed around the statue are analyzed by Feischmidt, *Ethnizität als Konstruktion* (fn. 19), 68-77. The monument was meant to become not only a landmark of the city, but also a national symbol.

any Hungarian or any German colonist.[21] To inscribe this idea into the public space, a copy of the Capitoline She-Wolf statue was brought to Cluj by an Italian delegation in 1921 and installed in the city's main square, just meters away from King Matthias' statue. The monument was destined to "proclaim the return of Transylvania to its origins and symbolize the *return of our rule* in this territory".[22]

The connection with Rome had been promoted in particular by the Romanian Greek-Catholics, a religious enclave that had emerged at the turn of the eighteenth century, when a part of the Orthodox clergy accepted Vienna's offer of a union with the Roman Church. The resulting access to higher education contributed to the emergence of the Greek-Catholic intellectual elite, which fully engaged itself into the Romanians' struggle for equal rights.[23] In Cluj only 11% of the population was Orthodox in 1930, while Greek-Catholics counted twice as much.[24] As such, their clergy equally planned to raise a cathedral in one of the city's main squares, and even addressed a request to the Municipality in this regard.[25] However, the idea was postponed until 1926, when the Vatican donated the centrally-located eighteenth century building of the Franciscan Monastery on their behalf.[26] While obtaining a suitable place for their ceremonies by adopting a practical approach, the Greek-Catholics failed to create an impact on the city's public space though a well-orchestrated display of national symbolism similar to that initiated by the Orthodox.

4. CONSTRUCTING THE CATHEDRAL BETWEEN NATIONALIST RHETORIC...

The construction of the Orthodox Cathedral in Cluj was linked in an almost organic manner to another similar project, the construction of the cathedral in Sibiu (Hermannstadt in German, Nagyszeben in Hungarian), built by the Romanian-Orthodox Bishopric in Transylvania at the turn of the twentieth century. Situated in southern Transylvania, close to the Romanian border, Sibiu had a German-speaking majority, but also one of the strongest ethnic Romanian communities in Transylvania. The construction of the Cathedral was financed exclusively from donations, and stood as a symbol for the emancipation of Orthodox Romanians. The construction process reflected the atmosphere of a multi-ethnic Empire: the declared loy-

21 This discourse started to be systematically formulated in the mid- eighteenth century by the Romanian Greek-Catholic intellectuals, becoming one of the main points in the argumentation for Romanians' emancipation. See Hitchins, Keith, *A Nation Affirmed. The Romanian National Movement in Transylvania, 1860-1914*, Bucharest 1999, 20-21.

22 *Înfrăţirea*, 30 September 1921.

23 Hitchins, *A Nation Discovered* (fn. 4).

24 *Recensământul general al populaţiei României 1930* ("The General Census of the population of Romania 1930"), vol. II, Bucureşti, 1930, XCIV.

25 In February 1922, the representatives of the Greek-Catholic Church addressed a similar request to local authorities. They were planning to construct a cathedral in the city center, and specifically indicated the Mihai Viteazul Square as a viable option. DJAN Cluj, Fond Primăria Municipiului Cluj-Napoca, Registre intrare-ieşire, I/ 262, filele 115-124.

26 *Patria*, 26 June 1926.

alty to the Emperor (Franz Joseph being also the first donator), the Byzantine design produced by Budapest-based architects,[27] the collaboration with workshops from all over the Monarchy and beyond and the speeches held in all the three main languages of Transylvania (i.e. German, Hungarian and Romanian) at the consecration festivities in 1906. Nevertheless, the limitations faced by the committee in charge with the construction works also showed the comparatively marginal position of the Romanians, as they could hardly obtain a more centrally-located parcel from the local government and were obliged to pull down their own parish church in order to make space for the new Cathedral.[28] In 1919, a group of Orthodox clerics led by Nicolae Ivan, financial councilor of the Orthodox Metropolitan Bishopric in Sibiu moved to Cluj to found a Bishopric for the northern part of Transylvania and also to build a cathedral.[29] Upon his arrival, Ivan met a situation when the small Orthodox church stood hidden in the city's periphery. For Ivan the urban topography illustrated the existing power relations between the ethnic groups and religious confessions in the city. The remoteness of the humble building seemed a humiliation to him. By contrast, even if Hungarians had lost influence in public administration, he thought that "their churches" (i.e. Catholic, Calvinist, Lutheran and Unitarian) had preserved most of their properties.[30]

Centrality and visibility were therefore essential for the location of the new Cathedral. The Bishop petitioned the municipality for an empty plot in the city center, either the park in front of the National Theatre, or the Union Square in the vicinity of the fifteenth century Gothic church.[31] The first option was ultimately chosen. The small park stood on the place of the demolished medieval city walls and was surrounded by important administrative and cultural buildings like the Palace of Justice, the Chamber of Commerce and the National Theatre, which had all been built around the turn of the twentieth century.[32] Yet, the initial reply of the Municipality was

27 Romanian Orthodox churches in Transylvania had traditionally been constructed from wood (only rarely from stone), being stylistically inspired by Gothic or Baroque. Only with the late nineteenth century Byzantine models came to be seen as the proper expression of Orthodoxy. This was partially due to the Viennese influence, where this style was promoted by architects such as Theophil Hansen, the author of the Parliament on Ringstraße. See Jula, *Arhitectura religioasă* (fn.6), 50-52.

28 Puşcariu, Ilarion, *Biserica Catedrală de la Mitropolia Ortodoxă Română din Sibiu. Istoricul zidirii 1857-1906* ("The Cathedral of the Romanian Metropolitan Bishopric in Sibiu: the History of its Building 1857-1906"), Sibiu 1908.

29 When Nicolae Ivan arrived in Cluj in October 1919 from Sibiu, he was already 64. As a counselor of the Orthodox Metropolitan Bishopric in Sibiu, he had been actively involved in the construction works of the Orthodox Cathedral. He had also been member of the City Council in Sibiu and from 1898 member of the Central Committee of ASTRA, the most important cultural association of Transylvanian Romanians. Convinced that the economic emancipation was equally important, he founded small banks, such as Iulia in Alba-Iulia, Vatra in Cluj (1907) and Lumina in Sibiu (1910). Stanca, Sebastian, *Episcopia ortodoxă română a Vadului, Feleacului şi Clujului (1919-1929)* ("The Romanian Orthodox Bishopric of Vad, Feleac and Cluj"), Cluj 1930, 51-53.

30 AMC (fn.9), Fond V-11-919, doc. 47-920, fila 5.

31 A similar undertaking occurred in Alba-Iulia, where the Coronation Church was constructed just across the street from the Catholic Cathedral. Popescu, *Le style national* (fn. 11), 212-213.

32 The central area in which the new cathedral was to be situated had passed through a radical transformation at the turn of the century (1880-1915), when the Hungarian authorities had sponsored the construction of numerous institutions, such as the Franz Joseph University, the University Clinics and Library, the Palace of Justice, the Chamber of Commerce and Industry, the Theater etc. Monumentality, historicism and an eclectic decorative language characterized the architectural landscape of the city center. Vais, Gheorghe, *Clujul eclectic* ("Eclectic Cluj"), Cluj-Napoca, 2009.

evasive. It seems that the Bishop's proposal initially lacked support from the mayor and the local council. However, although their agreement was obtained "after great persistence from our side", as one representative of the Orthodox Church mentioned, the commissions responsible for public sanitation and engineering within the Municipality strongly objected the project, arguing instead for the preservation of the small park.[33] The reasons behind this decision are not totally clear, although it can be inferred that ethnic and perhaps religious motivations played a role.[34] Only the intervention of central authorities in Bucharest finally pressured local administration into approving the request. „The construction of a Greek-Orthodox Romanian cathedral in Cluj represents a cultural, religious and national necessity. Therefore, we believe that it should be constructed in the very heart of the city"[35], emphasized a document of the Ministry of Interior. On 10 October 1920, the Cluj Municipality complied with the central orders and agreed to yield the park in front of the National Theatre to the Orthodox Bishopric.

In his correspondence with various state institutions, Ivan made use of nationalist rhetoric grounded in the multi-ethnic character of the city, in the hope of attracting funds for the Cathedral.[36] He argued that for the non-Romanians in the city, the image of the Romanian royal family and of the central government would be associated with this Cathedral as the King and government representatives would be bound to visit it during their travels to Cluj. In addition, he emphasized that this monument would commemorate the Unification in 1918, since its construction would be possible due to this event. Similarly to medieval cathedrals, which were supposedly built on places of martyrdom[37], the foundations of the Orthodox Cathedral were said to be lying "on the bones of the heroes and the martyrs sacrificed in the fights for our liberation"[38].

The architectural competition was opened in 1921 and attracted the interest of architects from Transylvania and Bucharest.[39] Yet, many of

33 Stanca, *Episcopia ortodoxă română* (fn. 29), 74; AMC (fn. 9), Fond II-23-920; Fond V-11-919, doc. Primăriei 1487-1920. The Mayor's final approval carries nothing from the nationalist rhetoric usually associated with the Cathedral. It simply states that a plot from Cuza Vod Square was yield to the Orthodox Bishopric for the construction of the Greek-Orthodox church in the hope that it would bring "moral benefit to Cluj inhabitants" and would contribute to "the beautification of the public space through the construction of a monumental building".

34 *Înfrăţirea*, 26 September 1920. One can assume that the request of the newly established Orthodox Bishopric in Cluj took local authorities by surprise, since this religious community was one of the smallest in the city. In the early 1900, the Greek-Catholics were eight times more numerous as compared to the Orthodox. DJAN Cluj, Fond ASTRA, Pachetul 2, fila 3. Iulian Pop, the first Romanian mayor of Cluj, was a Greek-Catholic. Also, the engineers working for these technical commissions were most probably Hungarians and Jews, since such specialists usually preserved their positions in the local administration due to their expertise (*Monitorul Municipiului Cluj*, no. 19, 1927).

35 AMC (fn. 9), Fond II-23-920, doc. 2650.

36 See for example the letter addressed to the Minister of Agriculture on 20 September 1920 in AMC (fn. 9), Fond V-11-919, doc. Ministerului Agriculturii si Domeniilor nr. 40348-1920.

37 Vauchez, "The Cathedral" (fn. 2), 41.

38 *Zile memorabile (4, 5 si 6 Noiembrie). Sfinţirea Catedralei Ortodoxe Române din Cluj* ("Memorable Days: (November 4, 5 and 6). The Consecration of the Romanian Orthodox Cathedral in Cluj"), Cluj 1933, 19. The cathedral was dedicated to the Assumption of the Virgin, celebrated on 15 August – incidentally the day when Romania entered the war in 1916.

39 Vasiu, Nicolae (ed.), *Episcopul Nicolae Ivan 1855-1936*, Cluj-Napoca 1985, 102. Among the architects who

these architects were known only locally, and, to the organizers' general disappointment, they arguably failed to capture the "essence" of Romanian Orthodoxy into their proposed designs, or did not meet the requirements for monumentality. The typical design expected by the organizers would reflect the characteristics of the Romanian national style, which had developed starting with the late nineteenth century, taking inspiration from the late medieval churches of Wallachia and Moldavia, as well as from vernacular and peasant architecture.[40] One of the surprising presences in the competition was Károly Kós, a local Transylvanian architect of Hungarian origin and a promoter of the folklore-inspired style, who chose the plan of a Late-Antiquity Byzantine church, influenced by the famous Hagia Sophia in Istanbul.[41]

The winning project was authored by the two Bucharest-based architects, George Cristinel and Constantin Pomponiu. Besides demonstrating a more clear understanding of the Romanian national style and its claimed Byzantine origins, the two architects wrapped their project in a nationalist rhetoric and declared that they got involved in this competition "animated by a patriotic feeling"[42]. The authors emphasized the dome above the nave as the distinctive characteristic of the building. In a very explicit manner, the architects explained that this cupola would be placed in the axis of Iuliu Maniu Street, which connected Cuza Vodă Square with the Union Square, where the Saint-Michael Gothic church was situated. Therefore, anyone walking on this street could see and compare the two churches.[43] The familiar cross-shaped plan with an emphasis on the longitudinal axis was used, while the lateral apses, although diminished, alluded to the traditional triconch specific to medieval Romanian churches. The element that was strangely "foreign" from the Byzantine context was precisely the dome, reminding of the Parisian Pantheon and Saint Paul's Cathedral in London rather than the traditional Byzantine church.[44] The Cathedral in Cluj was one of the

---

showed interest in the projects, the Bishop's correspondence mentions Károly Fényes (Arad), Dumitru Simu (Timişoara), Marcel Maller (Bucharest), and Karl Ballereich (Sibiu). AMC (fn. 9), Fond II-23-920, doc. 1742-921, 1854-921, 2118-921, 2274-921. Among those who submitted design proposals: Victor Vlad (Timişoara), Buermes and Strenzel (Sibiu), Ioan Pamfilie (Sibiu), Ioan Alexy and Zoltán Lothariu (Cluj) and Sterie Becu (Bucharest).The names are preserved in a larger envelope with the mention "Autorii planurilor înaintate pentru zidirea Catedralei" ("The authors of the plans submitted for the construction of the Cathedral").

40 Popescu, *Le style national roumain* (fn.11), 28-30. The development of the Romanian national style followed the emergence of Romania as a nation-state in the second half of the nineteenth century. The major role attributed to Byzantine art was due to its "discovery" by those architects conducting studies and restoration works of medieval churches (Popescu, *Le style national roumain* (fn.11), 20). During the interwar period, the Romanian national style became the image of the new state, its formal vocabulary being used for representative buildings in administration, education, and culture, as well as for residential quarters of the elites (Jula, *Arhitectura religioasă* (fn. 6), 39).

41 AMC (fn. 9), Fond II-23-920, doc. 3002-921, fila 1.

42 The Byzantine Empire was perceived as a source of political and religious power in the Balkans, and Romania, just like Serbia or Bulgaria, proudly considered itself as one of its legitimate heirs. In terms of church building, the Byzantine prototype presented a very practical advantage: it offered monumental scale models, which although missing in the medieval Romanian architecture, were necessary for the urban areas in which these cathedrals were to be placed. Ioan, *Power, Play and National Identity* (fn.11), 20.

43 AMC (fn. 9), Fond II-23-920, doc. 3588-921.

44 Popescu, *Le style national* (fn. 11), 255; Ioan, *Power, Play and National Identity* (fn. 11), 30. Stanca, *Episcopia ortodoxă română* (fn. 29), 77.

first religious buildings in Romania to make extensive usage of reinforced concrete, which was filled with bricks, and covered in stone, marble and ceramics.[45] By combining the architectural forms of medieval Byzantine churches with the advantages of modern construction technology [fig. 1], the Cathedral could be considered a form of "invented tradition"[46]. With a height of almost 70 meters, its elegant silhouette integrated well into the existing square, being surrounded by a small park.[47]

Although the Bishopric in Cluj claimed that the cost of the building would not exceed 12 million Lei, the completion of the winning project required a budget five times higher. If the central government provided most of the financial support, this was not due to a consistent state policy, but to the pressures exercised by local actors. The most active and enthusiast non-state promoter of the project was probably the Society of Orthodox Women, who strongly supported the construction of the Cathedral by organizing fund raising campaigns and patronizing a variety of cultural events, such as concerts, theatre plays, and conferences.[48] The economic crisis of the late 1920s also reflected on the progress of the construction works for the Cathedral. A postcard circulated in 1929 in order to collect additional funding

Fig. 1 Rising the cathedral's reinforced concrete skeleton.

45 Jula, *Arhitectura religioasă* (fn. 6), 126.

46 I use the concept from Hobsbawm, Eric/Ranger, Terrence (ed.), *The Invention of Tradition*, Cambridge 1992. Rather than properly traditions, monuments can be interpreted as visual expressions of prevailing visions of a forged collective identity, being associated with ceremonies and rituals conceived as invented traditions. For example, the architectural style of a building can reinterpret traditional morphological or decorative elements, thus connecting the present and the past in a symbolic visual discourse.

47 Moraru, Alexandru, *Catedrala Arhiepiscopiei Ortodoxe a Vadului, Feleacului şi Clujului* ("The Cathedral of the Orthodox Arch-Bishopric of Vad, Feleac and Cluj"), Cluj-Napoca 1998, 75.

48 *Renaşterea*, 6 January 1924. The name of the Society of Orthodox Women was inscribed on one of the four bells of the Cathedral, as a sign of gratitude for its activity. AMC (fn. 9), Fond II-8-923, doc. 6628-925.

dramatically pictured the skeleton of the cathedral rising in the cold winter. [fig. 2] Bishop Ivan complained about the lack of financial support from central and local authorities, claiming that the difficult situation had "forced us (i.e. the Orthodox Romanians) to stay for ten years now in this small parish church that can hardly accommodate 300 humble believers."[49] Local newspapers tried to influence both the public opinion and the government, by invoking sensitive issues such as the memory of the war: "This monumental work [...] constructed in the memory of those who died on the battlefields of the world war [...] does not belong to a parish or an eparchy, but to the liberated Transylvania", wrote the newspaper Patria in 1929.[50]

Fig. 2 The cathedral during the construction works in the late 1920s.

49 AMC (fn. 9), Fond II-23-920, doc. 2901-1922.

50 *Patria*, 1 May 1928. See also Moraru, *Catedrala* (fn. 47), 40-42.

5. ... AND THE PRACTICE OF MULTI-ETHNICITY

The multi-cultural environment in which the Orthodox Cathedral was built was well reflected by the participating companies and workshops. Rather than a product of Romanian national purity, the Cathedral was constructed by people of diverse ethnic backgrounds. When reading the documents preserved in the Archives of the Bishopric, it is interesting to notice that Nicolae Ivan attempted to continue the practices used during the construction works for the Sibiu Cathedral, for example a successful cooperation with specialists from the area of the former Austro-Hungarian Monarchy. The correspondence of Ivan was carried on mostly in Romanian, but Hungarian and German as well. The Bishop personally replied to every letter in the original language and also chose to order the Cathedral's four bells from the Hungarian town of Sopron to the Seltenhofer Workshop owned by a Jewish family (incidentally, the company had produced the bells for Sibiu cathedral as well).[51] Other works, such as the sculpture or the electrical installation, were entrusted to small workshops from Cluj.[52] Not least, the list of seasonal construction workers (bricklayers) included equally Hungarian and Romanian names.[53]

The Bishop made efforts to reduce the costs by asking for discounts for raw materials or transportation from various ministries.[54] The national rhetoric associated with the Cathedral was sometimes useful, but in many cases practical motivations prevailed and support was denied. In 1933, Patriarch Miron Cristea, a friend of Ivan, convinced King Carol II to donate the main chandelier for the Cathedral, in the shape of the royal crown, arguing that this gesture would symbolize that "the light comes from Bucharest."[55] On another instance, however, when Ivan asked for the wooden skeleton already used for the construction of the Coronation Church in Alba-Iulia[56], he was refused and the materials were send to Galaţi, a city affected at that time by a flood.[57]

Radical Romanian nationalists proved unsympathetic towards Ivan's vision of nationalism, and interpreted it as treason of national values. For example, a scandal was caused by the choice of the construction company. The decision of the Bishopric in favor of a company based in Cernăuţi (Cernowitz) caused negative reactions in the local press, as it was claimed that the construction of the Romanian Cathedral could not be entrusted to "non-Ro-

---

51 AMC (fn. 9), Fond II-8-923, doc. 5454-1925. The correspondence was usually carried in the language used by the sender. On the back of each letter he received, the Bishop wrote a draft answer. Although most of the Hungarian and Jewish companies wrote in Romanian, some used also Hungarian or German.

52 Theodor Orban and Vasile Roşca collaborated for the electrical installation (Ibid., doc. 4496-1933), Bauer and Nagy sculpted the stone decorations following the Byzantine tradition (Ibid., doc. 6741-1933).

53 Ibid., doc. 7598-1928. The workers' names appear at the end of a petition in which they complain about a deduction of their salaries, explaining they are seasonal workers that have to support their families from this money.

54 Ibid., doc. 1503-924. See for example this request to the National Railway Company, asking for a deduction of 20% from the transportation price for raw materials.

55 Letter written in Bucharest, on 8 July 1933, published by Vasiu, *Episcopul Nicolae Ivan* (fn.39), 248.

56 AMC (fn. 9), Fond II-23-920, doc. 874-923.

57 Ibid., doc. 1260-923.

manians," while this company was owned by two Germans and a Jew.[58] Following also financial disagreements with this company, Ivan was forced to reevaluated the options and choose a Bucharest-based company.[59] This time, another newspaper accused a politically-motivated decision, as the winning company was arguably supported by the Liberal Party in power. [60]

Various occasions related to the construction works, such as placing of the foundation stone in 1923, the inauguration of the bells in 1926, the consecration of the Cathedral in 1933 were used as opportunities for staging ceremonies in the city's public space.[61] These events included processions and speeches by officials, being attended by numerous dignitaries (the royal family, members of the government, local politicians, intellectuals, the Army), as well as by large crowds of peasants from the neighboring villages. The representatives of the other religious confessions were usually invited. These moments were used by Nicolae Ivan as means to attract public attention for the Cathedral – in order to show that the construction works were progressing and ask for additional funding, but also to promote Romanians' presence in the public sphere. Naturally, the most elaborated solemnity was dedicated to the Consecration of the Cathedral, which took place at the beginning of November 1933. It was a major event for the city, as the king and members of the government were present. The ceremony in front of the Cathedral was orchestrated with the use of modern technology; the megaphones made the ceremony be heard in the entire city center. [62]

As the major construction site in the city center, the rising of the Romanian Orthodox Cathedral naturally attracted public attention. The two major events – the beginning and the end of the construction works – received generous coverage in the most important local Hungarian and Jewish newspapers. The reports did not limit themselves to a detailed presentation of the facts, but also seemed to display a conciliatory attitude and acknowledged the Orthodox Romanians' right of having a representative building constructed in the city center. The construction of the Romanian cathedral in Cluj was definitely perceived as a symbol of change, a building which is "marking and making history", as one Hungarian journalist put it.[63] Similar articles referred to a spirit of community and tolerance that should characterize the city residents, and subtly tried to bring in the minorities' own requests in terms of language and religious equality. However, beyond such rhetoric, it appears obvious that Hungarian and Jewish newspapers associated peculiar political connotations to the Romanian cathedral, in which the presence of the King definitely played a major role.[64] Moreover, the

58 *Clujul*, 17 June 1923.

59 Stanca, *Episcopia ortodoxă română* (fn. 29), 78-79.

60 *Patria*, 20 June 1923.

61 The first of these ceremonies was the laying of the founding stone, which took place on 7 October 1923. (*Patria*, 5 October 1923, *Renaşterea* 14 October 1923), followed by the consecration of the main cross on 6 July 1926 (*Renaşterea*, 15 August 1926) and the consecration of the bells (*Renaşterea*, 24 April 1927).

62 AMC (fn. 9), Fond II-23-920, doc. 2650, 10.

63 *Keleti Újság*, 7 October 1923.

64 *Keleti Újság*, 9 October 1923; *Keleti Újság*, 5 November and 7 November 1933; *Új Kelet*, 5 November and 7 November 1933.

presence of a non-urban element – thousands of peasants brought from the neighboring villages to attend the ceremonies – created a certain discomfort among the Hungarian-speaking townsmen, who resented their presence as something invasive.

The Calvinist Bishop in particular seemed to be in excellent terms with Bishop Ivan, publishing a congratulatory article in his honor.[65] On the other hand, Romanian Greek-Catholics, although sympathetic towards the national and political significance of the new monument, felt that the actual costs of the project was also of importance. Their official newspaper, *Unirea*, described the event as a ceremony that had received national proportions due to the participation of the King and the political elite. The Cathedral could indeed be seen a national symbol; yet the government had paid more than 50 million Lei from the state budget for its construction. [66] In that way, Greek-Catholics suggested, the new Orthodox Cathedral was constructed by donations of all Romanian taxpayers, irrespective of confession or ethnicity.

6. CONCLUSION

In 1936, the year of Bishop Ivan's death, the Orthodox Cathedral in Cluj was standing proud in the middle of a new square [fig. 3], just a few hundred meters away from the city's older churches. For the Romanian elites contrib-

Fig. 3 The Orthodox Cathedral in Cluj upon its Completion, in the late 1930s.

65 *Református Szemle*, nr. 30-31 (1933), republished in *Zile memorabile* (fn. 34), 96.

66 *Unirea*, 11 November 1933.

uting to its construction, the monument created a sense of historical justice, a symbol of their rightful place in the city's hierarchies of power. The traditional architectural forms covering the reinforced concrete skeleton, inserted in the midst of the Baroque and historicist buildings, portrayed Transylvanian Romanians' own claims to historical continuity, which they could now inscribe into the ideology of a modern nation-state. From beginning to the end, the construction process has been undeniably wrapped in the political discourse of nationalism. Yet, in ways previously experienced in the context of the Austro-Hungarian Empire, nationalist rhetoric was combined with negotiation in the everyday practices of a multi-ethnic city. Ultimately, the actors involved in the construction site of the Orthodox Cathedral were not only building a religious edifice, but they were also creating ambivalent tactics of promoting their own agendas by exploring the ambiguities embedded in this different political context.

LIST OF FIGURES

Fig. 1: Biblioteca Judeţeană "Octavian Goga" Cluj-Napoca, Memorie şi Cunoaştere Locală ("Octavian Goga"District Library, Section for Local Memory and Knowledge).

Fig. 2: Biblioteca Judeţeană "Octavian Goga" Cluj-Napoca, Memorie şi Cunoaştere Locală ("Octavian Goga" District Library Cluj, Local Memory and Knowledge Section) (online at http://www.bjc.ro/wiki/index.php/Imagine:Catedrala-ortodoxa_2.jpg, accessed 7-4-14).

Fig. 3: Biblioteca Judeţeană "Octavian Goga" Cluj-Napoca ("Octavian Goga" District Library Cluj, Local Memory and Knowledge Section) (online at http://www.bjc.ro/wiki/index.php/Imagine:Catedrala-ortodoxa.jpg, accessed 7-4-14).

CONSTRUCTING A DAM-NATION?
CONSIDERATIONS ON THE ROGUN DAM IN TAJIKISTAN[1]

Filippo Menga

1. INTRODUCTION

More than 45.000 large dams were built around the world during the twentieth century, obstructing two thirds of all freshwater flows[2] and harnessing water resources for food production, energy generation, flood control and domestic use.[3] Dams became, to many, synonymous with development and progress, representing modernity and humanity's ability to tame water resources. Major dams,[4] in particular – that are among the largest structures built by humans – not only physically alter the landscape but also shape perceptions and ideas as they symbolize the might of the state that built them, and thus they often become a favourite of nation-builders around the world.[5]

While the number of dams being erected worldwide declined from the 1970s onwards due to the emergence of the environmental awareness paradigm inspired by the green movement,[6] dams are now back on the global agenda. Hundreds of new, controversial and extremely costly projects have been launched in the last few years, sided by the promise of prosperity and overly optimistic predictions being made by project promoters.[7] In this novel dam frenzy, China, Brazil and India have taken the role that once belonged to the Soviet Union (USSR);[8] guided by Lenin's motto "Communism is Soviet power plus the electrification of the whole country"[9], the Soviets had indeed built some of the largest hydroelectric dams in the world (such as the Sayano–Shushenskaya and Dnepr Dam in Russia, or the Nurek Dam in Tajikistan).

---

1 This chapter is an elaborated version of an article published in Nationalities Papers (Menga, Filippo, "Building a nation through a dam: the case of Rogun in Tajikistan", *Nationalities Papers*, vol. 43, 2015, 479-494).

2 Baghel, Ravi/Nüsser, Marcus, "Discussing Large Dams in Asia after the World Commission on Dams. Is a Political Ecology Approach the Way Forward?", in: *Water Alternatives*, vol. 3, 2010, 231-248.

3 World Commission on Dams, Dams and development, *A new framework for decision-making. The report of the World Commission on Dams*, London, Sterling, 2000.

4 The International Commission on Large Dams (ICOLD) defines a major dam as a dam with a height of 150 meters or more from the foundation, a reservoir storage capacity of at least 25 cubic kilometers and an electrical generation capacity of at least 1000 megawatt.

5 McCully, Patrick, *Silenced Rivers. The Ecology and Politics of large Dams*, London et. al. 2001.

6 Allan, Tony, "IWRM/IWRAM: a new sanctioned discourse?", in: *SOAS Occasional Paper* 50, April 2003.

7 Ansar et al., "Should we build more large dams? The actual costs of hydropower megaproject development", in: *Energy Policy*, vol. 69, 2014, 43-56.

8 Gleick, Peter, *The World's Water (Volume 7): The Biennial Report on Freshwater Resources*, Washington D.C., 2011.

9 Lenin, Vladimir Ilyich, "Our Foreign and Domestic Position and Party Tasks. Speech Delivered To The Moscow Gubernia Conference Of The R.C.P.(B.)", in: *Lenin's Collected Works*, Moscow [4]1965, 408-426.

This article takes as a case study the revitalisation of one of these Soviet projects, the massive Rogun Dam on the Vakhsh[10] river in Tajikistan, to outline the discursive constructions that can accompany the realisation of large hydraulic infrastructures. Although there might be a seeming continuity between the rhetoric adopted by the Tajik government and Maxim Gorki's socialist ideological production of the 1920s and 1930s,[11] a key difference exists between the two. While the Soviet administration considered large-scale construction projects as means to an end (with the end being technological modernization and economic self-sufficiency), for the Tajik government the construction of the Rogun dam has arguably become the end in itself.

Through the analysis of the speeches delivered by the President of Tajikistan Emomali Rahmon and his ministers at national and international summits, the present article discusses how the construction of a large dam can be used by a ruling political elite to construct a national ideology, outlining the profound link between the performative (e.g., variations in the water flow, preventing floods, allowing irrigated agriculture) and discursive effects that come along with the construction of a large dam. Indeed, through an effort aimed at persuading its citizens and the international community that the construction of Rogun is a fundamental element in the country's nation-building process, the Government of Tajikistan (GoT) has created what can be defined a "Rogun ideology".

This aspect of persuasion, in particular, is relevant to the study of how the building of a dam can become a symbol of success and national cohesion, and of how its significance is discursively constructed by a ruling elite. In this regard, useful insights to a deeper understanding of this subject are offered by the Foucauldian concept of normalization – intended as the social process through which the compliance to certain ideas, norms and behaviours is gradually seen as normal[12] – and perhaps even more prominently by the Gramscian concept of hegemony. The latter denotes indeed the success of a dominant class in presenting its view of the world and its ideology in a way that the other classes accept it and consider it common sense.[13] Although coercion and consent come together, and they are, in the function of hegemony, "connective"[14], it is primarily on consent that a State needs to base its relations with the civil society. If the State is able to create in people certain expectations and behaviours, and shape their ideology, it will be successful in affirming and consolidating hegemony. Ideologies, for Gramsci, are assessed for their social effects rather than on their effective value.[15] Based on these assumptions, the rhetorical strategies adopted by the Tajik government can be seen as part of a process aimed at creating and affirming a specific "dam-ideology", one that can help it legitimize its actions while gaining popular support and consent.

10 The Vakhsh river is one of the main tributaries of the Amu Darya river, the largest river of Central Asia.

11 See for instance, in this volume, Marie Collier's discussion of the rhetorical representations of construction sites in the Soviet Union presented in the periodicals *Nashi Dostizheniya* and *SSSR na Stroike*.

12 Foucault, Michel, *Discipline and punish. The birth of the prison*, York ²1995.

13 Gramsci, Antonio, *Quaderni del carcere. Edizione critica dell'Istituto Gramsci*, Torino 1975.

14 Ibid., Q12 §1.

15 Fairclough, Norman, *Critical Discourse Analysis. The critical Study of Language*, London 2010.

2. LINKING NATION-BUILDING TO DAM-BUILDING

While nationalism studies have been marked by a central and sharp debate about how and when nations first appeared, the importance of symbols in the creation of national identities has been a less controversial matter. For instance, perennialist theorist Anthony D. Smith observed that "memories, myths, symbols and values [...] furnish a distinctive and varied repertoire from which different elites can select those elements which can mobilize and motivate large numbers of their designated population"[16]. Likewise, the modernist Ernest Gellner also acknowledged the crucial role of symbols and symbolism in the study of nations and nationalism.[17] Following the constructivist paradigm to nation-building, this study considers nations and nationalism as social constructs resulting from different discourses and narratives, adopting Benedict Anderson's conception of national identities as cultural artefacts and the nation as "an imagined political community"[18]. In this context, ruling elites[19] can image and construct national identities by consciously manipulating myths and symbols and producing a specific nationhood narrative, using for instance the symbolism stemming from the construction of large-scale infrastructures.[20]

Indeed, elite nationalism often overlaps with megaprojects, whose construction can be interpreted as the imposition of a dominant perspective. As Sanjay Sangaval effectively explains, "by fashioning nation and nationalism to include only the dominant classes and their interests, protests against mega projects [...] become by definition 'anti-national' and the people who protest [...] anti-nationals"[21]. Politicians thus present the realisation of new projects such as roads, rail links, bridges or stadiums as key national achievements, largely overlooking their costs and inflating their benefits, through what Bent Flyvbjerg has defined a "mega delusional" process.[22]

Among the various types of megaprojects, large dams, which bring with them the ability to dominate nature and use its power to serve the needs of society, have long fascinated politicians and rulers. Water, that can be diverted and used to irrigate arid areas and thus sustain livelihoods, offers a good example of how natural resources can assume a central role in the development of a civilization. Powerful ancient empires, such as the Chinese,

16 Smith, Anthony D., "LSE Centennial Lecture. The Resurgence of Nationalism? Myth and Memory in the Renewal of Nations", in: *The British Journal of Sociology*, vol. 47, 1996, 575-598, 591.

17 Gellner, Ernest, "Ernest Gellner's reply: 'Do nations have navels?'", in: *Nations and Nationalism*, vol. 2, 1996, 366–370.

18 Anderson, Benedict R., *Imagined Communities*, London, 32006, 6.

19 For the purposes of this article, the term denotes "people with attributes that lead them to be ranked higher and accorded more prestige and respect than ordinary people", adopting the definition given by Withmeyer, Joseph, "Elites and popular nationalism", in: *British Journal of Sociology*, vol. 53, 2002, 321-341, 322.

20 It is worth noting, however, that nationalism does not necessarily stem from the elite, as there can also be cases of bottom-up popular nationalism, as for instance the one that emerged in China in the twentieth century (Gries, Peter Hays, "China and Chinese Nationalism", in: Delanty, Gerard/Kumar, Krishan (ed.), *The SAGE Handbook of Nations and Nationalism*, Thousand Oaks 2006, 488-499).

21 Sangval, Sanjay, "Nation, 'Nationalism' and Mega Projects", in: *Economic and Political Weekly*, vol. 29, issue 10, 1994, 537.

22 Flyvbjerg, Bent, "Mega Delusional. The Curse of the Megaproject", in: *New Scientist*, December 2013, 28-29.

Mesopotamian, Egyptian or Maya, used rivers to develop large-scale irrigated areas which contributed to their growth and expansion.[23] Nevertheless, if water can create, it can also destroy and, for instance, the idea of annihilating the rival city of Pisa through the diversion of the Arno River in Italy, was not such a remote option for sixteenth century Florence.[24] Hence, the connection between control of water and control of political power has long been studied. In this regard, a remarkable contribution was given by Karl August Wittfogel, who introduced the concepts of hydraulic society and hydraulic despotism in his provocative theory of *Oriental Despotism*.[25] According to Wittfogel, those who control water in arid or semi-arid regions also control political power; hydraulic regimes can increase their grip on power through the construction and management of hydraulic infrastructures such as dams and network of canals, that allow an elite of bureaucrats to wield control over their people and rivers. Building up on this, other scholars have studied how a ruling political elite can increase its power and preserve the existing social order through large hydraulic projects, as in the case of hydraulic development in the West of the United States during the end of the nineteenth and twentieth century.[26] In this period, and in conjunction with industrial modernity, the so-called hydraulic mission of states took place, reaching its peak in the mid-twentieth century both in liberal western economies and in the centrally planned economies of the Soviet Union. Subsequently, in the second half of the twentieth century, the hydraulic mission was also launched in what at the time was called the Third World,[27] clogging with dams most of the world's river basins.

It is in this context that some of the largest and most symbolic dam projects were realised. From Europe to America, from Asia to Africa, big dams have been held up as symbols of progress and patriotism. A key example is provided by the gigantic Aswan High Dam, which was completed in Egypt in 1971 with Soviet support, becoming "the centrepiece of postwar nation making"[28] in the country. John Waterbury observes that as relations between Egypt and Britain deteriorated in the 1950's, "Nasser and his associates could no longer regard the dam as simply a big engineering project, but rather came to hold it up as the symbol of Egypt's will to resist imperialist endeavours to destroy the revolution"[29]. If on the one side those who supported the Aswan High Dam were treated as patriots, on the other side, those who criticized it were "thought of as subversive or even treasonous"[30].

23 See for instance Wittfogel, Karl August, *Oriental Despotism: A Comparative Study of Total Power*, New Haven, 1957; Molle, Francois/Mollinga, Peter/Wester, Philippus, "Hydraulic bureaucracies and the hydraulic mission: Flows of water, flows of power", in: *Water Alternatives*, vol. 2, 2009, 328-349.

24 Masters, Roger D., *Machiavelli, Leonardo, and the science of power*, Notre Dame 1996.

25 Wittfogel, *Oriental Despotism* (fn. 23).

26 Worster, Donald, *Rivers of empire. Water, Aridity, and the Growth of the American West*, New York 1985; Reisner, Marc, *Cadillac desert. The American West and its disappearing Water*, New York 1993; Swyngedouw, Erik, "Modernity and Hybridity. Nature, Regeneracionismo, and the Production of the Spanish Waterscape, 1890–1930", in: *Annals of the Association of American Geographers*, vol. 89, 1999, 443–465.

27 Allan, "IWRM" (fn. 6), 10.

28 Mitchell, Timothy, *Rule of experts. Egypt, Techno-Politics, Modernity*, Berkeley, 2002, 45.

29 Waterbury, John, *Hydropolitics of the Nile Valley*, Syracuse 1979, 108.

30 Ibid., 117.

More recently, but still in line with the usual rhetoric associated to earlier dams, Chinese leaders portrayed the mastodontic Three Gorges Dam on the Yangtze River as the "greatest engineering feat since the construction of the Great Wall"[31]. Commenting on the Chinese dam, a report released by Human Rights Watch in the 1990s noted how Chinese Premier Li Peng seemed "intent on denying a public forum to opponents of the monumental project and on forcing it through both as a means of symbolizing China's fast-emerging "superpower" status and as a vehicle for personal glorification".[32] Finally, the Grand Ethiopian Renaissance Dam on the Blue Nile – as its name already anticipates – is also being depicted by the government as a unifying element for Ethiopians at home and abroad.[33]

For what concerns the Soviet Union, its hydraulic mission started in the late 1920s and continued until the late 1980s, when it was terminated by General Secretary Gorbachev in the dramatic political and economic circumstances of the period.[34] Driven by the desire of turning "mad rivers sane"[35], Soviet administrators pursued mostly two objectives through their hydraulic mission: increase agricultural and electricity production, through respectively large-scale irrigation projects and massive hydropower plants. In Central Asia, the hydraulic mission engendered the construction of large dams and water reservoirs in the mountainous areas of the upstream republics (Kyrgyzstan and Tajikistan) which, together with a complex network of canals, made it possible to practice irrigated agriculture in the plains of the downstream countries (Kazakhstan, Turkmenistan and Uzbekistan), where water intensive agricultural crops such as cotton, rice, and wheat were grown.[36] Moreover, through the construction of dams and canals, the Soviet administrators created a situation that would ensure competition between upstream and downstream countries, thus reinforcing the national distinctiveness of the republics and maintaining a role for Moscow as a dispute settler.[37] The first major irrigation projects began in 1939, with the construction of 45 canals, including the Great, the North and the South Ferghana canals in the Ferghana Valley.[38] Thanks also to the

31 Kaika, Maria, "Dams as Symbols of Modernization. The Urbanization of Nature between geographical Imagination and Materiality", in: *Annals of the Association of American Geographers*, vol. 96, 2006, 276-301.

32 Human Rights Watch, *The Three Gorges Dam in China: Forced Resettlement, Suppression of Dissent and Labor Rights Concerns*, 1995, 3.

33 Ministry of Foreign Affairs of Ethiopia, *The hostile campaign against GERD: more old wine in new bottles*, http://www.ethiofact.com/index.php?option=com_content&view=article&id=3495:the-hostile-campaign-against-gerd-more-old-wine-in-new-bottles&catid=125:ethiopianews&Itemid=321 (accessed 5-9-13).

34 Allan, Tony, "Water resource development and the environment in the 20th century: first the taking, then the putting back, The Basis of Civilization - Water Science?", in: *Proceedings of the UNESCO/IAHS/IWHA symposium held in Rome in December 2003*.

35 Maxim Gorky quoted in McCully, *Silenced Rivers* (fn. 5), 17.

36 Rakhmatullaev, Shavkat et al., "Facts and Perspectives of Water Reservoirs in Central Asia. A Special Focus on Uzbekistan", in: *Water*, vol. 2, 2010, 307–320.

37 O'Hara, Sarah, "Central Asia's Water Resources: Contemporary and Future Management Issues", in: *Water Resources Development*, vol. 16, 2000, 423–441.

38 Matley, Ian Murray, "Agricultural Development", in: Allworth, E., (ed.), *Central Asia, A Century of Russian Rule*, New York 1967, 266-308.

momentum gained with the Virgin Lands Campaign (launched in the 1950s by Krushev), the total irrigated area in Central Asia increased from 4.5 million hectares in 1965 to 7 million hectares in 1991.[39] During the same period, the largest Central Asian hydro-electric dams were designed and built, including the Toktogul dam on the Naryn River in Kyrgyzstan, and the Nurek dam in Tajikistan (which is at present the country's main source of electricity), the latter being part of a cascade system of eight dams on the Vakhsh River.[40] Along with these projects, the Soviet administration sought to block the Vakhsh river with another larger dam, Rogun, located about 70 km upstream of the Nurek reservoir. The idea of taming the Vakhsh River on top of the Nurek cascade system, fitted the overarching Soviet ideology of "technically, everything is possible"[41].

Originally conceived as a dual-purpose structure for irrigation water management and for hydroelectricity, the Rogun dam was designed in the 1960s by the Soviet Hydroproject Institute in Tashkent, which also carried out a first feasibility study. The original project consists of a 335-meter high structure, a reservoir with a volume of 13.3 km^3 and six 600 megawatts (MW) turbines, resulting in a total installed capacity of 3,600 MW.[42] If compared with other dams, Rogun would be the tallest in the world – the fourth one being Nurek (300-meter) – and the twentieth for installed capacity.[43] Preparatory construction works began in 1976, and intense construction started in 1982, involving five to ten thousand people.[44] In 1991, due to the collapse of the Soviet Union and the worsening political situation in Tajikistan, that would eventually lead to a five year civil war (1992-1997), works at the Rogun site were stopped. Furthermore, in 1993 – which was originally the year set for its first unit to start producing electricity[45] – the upper coffer-dam was washed away by a powerful flash-flood. Combined with inadequate management caused by the civil war, the flood destroyed most of the accomplished structure,[46] frustrating two decades of efforts and an investment of 802 million dollars, leaving the "Queen of the Tajik mountains without a crown"[47]. Nevertheless, the idea of building Rogun was already too well-established in the minds of Tajik bureaucrats to be washed away with the flood.

39 Wegerich, Kai, "Hydro-hegemony in the Amu Darya Basin", in: *Water Policy*, vol. 10, 2008, 71–88.

40 Libert, Bo et al., "Water and Energy Crisis in Central Asia", in: *China and Eurasia Forum Quarterly*, vol. 6, 2008, 9-20.

41 Eshchanov, Bahtiyor et al., "Rogun Dam – Path to Energy Independence or Security Threat?", *Sustainability*, vol. 3, 2011, 1578.

42 Schmidt, Roland, "Feasibility Study for Completion of the Rogun Scheme, Tajikistan", *Hydropower & Dams*, vol. 3, 2007.

43 International Commission on Large Dams, *Register of Dams*, http://www.icold-cigb.net/GB/World_register/general_synthesis.asp?IDA=207 (accessed 16-1-13).

44 United Nations Environment Programme, *Environment and Security in the Amu Darya Basin*, 2011, 48.

45 Yerofeyeva Natalya, "Rogunskaya GES v Tadzhikistane budet dostroyena. No dlya etogo nuzhny inostrannyye investitsii", in: *Rossiyskaya Gazeta*, 25 October 2002.

46 Fradchuk, Artyom, "Tajikistan's Energy Dilemma", in: *Institute for War and Peace Reporting*, http://iwpr.net/report-news/tajikistans-energy-dilemma (accessed 12-3-2012).

47 Dyuzheva, Irina, "Chto Nama stoit GES postroit", in: *Parlamentskie Gazeta*, http://old.pnp.ru/archive/10821742.html (accessed 8-1-2013).

3. INDEPENDENT TAJIKISTAN AND THE ROGUN DAM

Although Rogun used to be a Soviet project, its significance increased when the Soviet Union ceased to exist. With independence – and with the vanishing of the centralised Soviet management system responsible for the allocation of resources to the Soviet republics – energy-poor Tajikistan [fig. 1] had to start paying for the imports of gas, oil and coal necessary to fulfil its energy needs. However, the country's failure to pay for outstanding debts combined with a tense relationship with Uzbekistan (its sole supplier of natural gas), resulted in recurrent cuts in energy supplies and in frequent energy crisis. Around 70 per cent of the Tajik population experiences extensive electricity shortages in winter, which, alongside their social costs, cause economic losses estimated at over US$ 200 million per year.[48] Under such circumstances, the potential impact of a hydroelectric power plant of the size of Rogun seems remarkable. Namely, as Tajikistan's electricity production from hydroelectric sources accounts for around 97% of total,[49] the country's installed capacity of 4,000 MW could almost double with the additional 3,600 MW that the Rogun dam will generate, allowing Tajikistan not only to become energy secure but also to sell electricity to Afghanistan and Pakistan. This is even more relevant after the two exceptionally cold winters that hit Central Asia between 2007 and 2009, and that engendered a major energy crisis in Tajikistan and in Kyrgyzstan, which was further aggravated by the Kazak and Uzbek withdrawal from the Central Asia Power System.[50]

Nevertheless, it is perhaps at the political level that Rogun can have an even stronger impact. The collapse of the Soviet Union implied that Communism was no longer providing a basis for legitimacy to national

Fig. 1 Tajikistan Map.

48 The World Bank, *Tajikistan's Winter Energy Crisis. Electricity Supply and Demand Alternatives*, Washington, 2012, i.

49 The World Bank, *Electricity production from hydroelectric sources (% of total)*, http://data.worldbank.org/indicator/EG.ELC.HYRO.ZS (accessed 12-1-13).

50 The World Bank, *Tajikistan's Winter Energy Crisis* (fn. 48), 56.

governments, and this led former Communist leaders to take a nationalist turn to enhance the perceived legitimacy of their authority.[51] The symbolism and prestige that can be attached to the world's tallest dam can be used by the Tajik government to disseminate a specific rhetoric aimed at getting legitimation, gaining consensus and diverting attention from more pressing matters. Even more so, considering that not long ago Tajikistan – the least prepared of the Central Asian countries to undergo policies of national consolidation[52] – was ravaged by a harsh civil war (1992-1997) that enfeebled the authority of the national government and accentuated regional and clan divisions.[53] The unifying effect of an iconic project like Rogun can contribute to the creation of a national identity, while helping keep in power President Rahmon and his close network from the Kulob region.

It is also worth noting that Tajikistan – along with the other Central Asian countries – does not seem at the forefront in the promotion and application of the general principles of accountability and good governance. The country can be defined as a neopatrimonial regime, whose apparent democratic structure is relying on clientelism, corruption, and personalistic rule to guarantee continuity in power.[54] This is even more relevant to this analysis, if we consider that the hydropower sector is frequently linked with corruption, and by many is actually classified as the most corrupted industrial sector.[55] For instance, Transparency International, an NGO which monitors corporate and political corruption, dedicated its 2008 Global Corruption Report to "Corruption in the Water Sector", observing that:

> Of the US$11.1 trillion the world is predicted to spend on energy infrastructure between 2005 and 2030, US$1.9 trillion may be expected to go toward hydropower. These large numbers create multiple opportunities for bribery, fraud and other forms of corrupt behaviour. [...] Combined with a lack of transparency, this provides fertile ground for manipulation and abuse.[56]

Therefore, besides the political and symbolic value of the dam stemming from a fascination of scale and even to a certain *folie de grandeur*[57], the less idealistic dimension related to corruption and to the redistribution of wealth among the elite cannot be completely overlooked.

---

51 Mellon, James G., "Myth, Legitimacy and Nationalism in Central Asia", in: *Ethnopolitics*, vol. 9, 2010, 137-150.

52 Gleason, Gregory, *The Central Asian States. Discovering Independence*, Boulder 1997.

53 Akiner, Shirin, *Tajikistan. Disintegration or reconciliation?*, London 2001.

54 Sehring, Jenniver, "Path dependencies and institutional bricolage in post Soviet water governance", in: *Water Alternatives*, vol. 2, 2009, 61-81.

55 McCully, *Silenced* (fn. 5); Pearce, Fred, *When the Rivers Run Dry: Water. The Defining Crisis of the Twenty-first Century*, London 2007.

56 Transparency International, *Global Corruption Report 2008: Corruption in the Water Sector*, New York 2008, 86-87.

57 For instance, in 2011 Tajikistan inaugurated the world's tallest flagpole (165 meter), marking the twentieth anniversary of independence and taking a record that before belonged to Turkmenistan. During the ceremony, the Tajik President hailed the huge flag as a "sign of independence, one of the main symbols of statehood and national unity [...] the sign of the struggle of the Tajik people, the lives they have sacrificed for freedom and independence, as well as the sign inspiring faithful sons of the nation to selfless work and activities and guiding the people of our country towards prosperity and flourishing and happy life" (Tajik President inaugurates "world's tallest" flagpole, *BBC Monitoring Central Asia Unit*, 30 August 2011).

Furthermore, along with its benefits, the dam is also bringing a few downsides. To begin with, Rogun has caused the deterioration of Tajik relations with downstream countries, and in particular with Uzbekistan. This is caused by concerns on a possible shortage of irrigation and drinking water[58] in summer due to winter release for electricity generation, in conjunction with fears that the seismicity of the area where the project is being built could lead to environmental calamities. The usual anti-Rogun rhetoric is well summarised by the declarations of Uzbek President Islam Karimov. In 2010, when asked why Uzbekistan is opposing the construction of Rogun, he stated "How can we let the residents of Uzbekistan live without water for eight years, while the Rogun water reservoir is being filled up? What will farmers be doing all this time?"[59] A few weeks earlier, while addressing the Plenary Session of the UN Millennium Development Goals (MDGs) Summit, Karimov made an implicit reference to Rogun when he declared:

> And in these conditions any attempts to implement projects drafted 30-40 years ago, yet in the Soviet period, to construct in the upper stream of these rivers [the Amu Darya and Syr Darya] the large scale hydropower facilities with gigantic dams, and moreover, if to take into account that the seismicity of the area of forthcoming construction makes up 8-9 points, - all of these may inflict an irreparable damage to environment and will be a reason for the most dangerous man-caused catastrophes which we have been witnessing for over the last years. As many international ecological organizations and respected experts recommend, it would be much more rational to switch to building less dangerous, but more economical small Hydropower Stations to have on these rivers the same energy power generating capacities.[60]

As tensions escalated, in 2012 Karimov went as far as saying that a project like Rogun could lead to "not just serious confrontation, but even wars"[61]. In response to these criticisms, the GoT has reassured its neighbours that Rogun will be operated in the interest of all basin riparians, adding that it will nevertheless proceed with the construction of the dam. The project has raised complaints within Tajikistan itself, due to the foreseen forced resettlement of the 30,000 people living in the Rogun, Nurobod and Rasht areas,[62] but also in this case the GoT was not deterred in its intentions.

There is, however, another obstacle to the construction of Rogun, one that cannot be easily overcome: the project is extremely expensive. With a total cost of US$ 2.9 billion, it cannot be financed by Tajik national resources alone. Although the GoT has calculated that US$ 800 million of work has already been executed, the project still requires US$ 2.1 billion of funds,[63]

58 Rogun could impact on the water flow of the Amu Darya River – of which the Vakhsh is one of the main tributaries – which flows from Tajikistan to Afghanistan, Turkmenistan and Uzbekistan. The riparians of the Amu Darya River Basin are Tajikistan, Afghanistan, Turkmenistan and Uzbekistan.

59 Rogun project undermines Uzbekistan's water supplies – Karimov, *Interfax News Agency*, 12 October 2010.

60 Karimov, Islam, "Address by H.E. Mr. Islam Karimov, President of the Republic of Uzbekistan, at the Plenary Session of the UN Millennium Development Goals Summit, New York", 20 September 2010, 3.

61 Karimov, Islam, "Water Wars in Central Asia. Dammed if they do", in: *The Economist*, 29 September 2012.

62 Bureau of Human Rights and Rule of Law, *Report on relocation cases study results*, Dushanbe 2012.

63 Eurasian Development Bank, *Water and Energy Resources in Central Asia. Utilization and Development Issues, Industry report*, 2008, 20.

equivalent to roughly a third of the country's 2011 GDP.[64] In order to meet this necessity of external funding, the GoT has carried out an uninterrupted effort over the last twenty years aimed at the mobilisation of financial resources. Significantly, a treaty was signed in October 2004 by Rahmon and his Russian counterpart Vladimir Putin. One of its effects was that the Russian aluminium giant RusAl agreed to invest US$ 560 million to resume work at the Rogun site and complete the construction of the first stage of the project.[65] The German engineering firm Lahmeyer, which was awarded a contract from RusAl to carry out a first feasibility study of the dam,[66] recommended 285 meters as the ideal height of the dam, instead of 335, on which the GoT insisted.[67] Without these additional 50 meters, Rogun would still be a very tall dam, but not the tallest in the world. Such was the disagreement on the height of the dam that, after an initial stall in construction works, in August 2007 Tajik officials eventually announced the cancelation of the deal with RusAl.[68] This insistence on having the tallest dam in the world seems of interest, especially if Rogun is analysed through the lenses of its symbolic significance. If, indeed, the GoT wants to persuade its citizens that the dam is a symbol of national pride and success, the power of suggestion that derives from a structure that stands taller than any other, probably helps in making this operation successful.

As a consequence of the Uzbek reiterated requests of having an external examination of the project, the World Bank (WB) got involved in the matter. In 2010, after a round of consultations with the riparian countries that went on from October 2008 until April 2009, Motoo Konishi, the WB regional director for Central Asia, announced that the bank will carry on an 18 month feasibility study and environmental assessment of the dam.[69] More precisely, a Techno-Economic Assessment Study (TEAS) and an Environmental and Social Impact Assessment (ESIA) were contracted respectively to a consortium led by Coyne & Bellier and to the company Poyry of Switzerland. In the meanwhile, construction works – that until then advanced intermittently and postponing major river blockages – were stopped in 2012. As of June 2014, the feasibility study is yet to be released,[70] and the Tajik government agreed that "no new construction would commence until the Assessment Studies have been prepared, reviewed by the Panels of Experts, then shared and discussed with riparian nations"[71].

With all the above in mind, the main drivers of the GoT's mission to popularize the Rogun dam seem to be the will to create a national

64 The World Bank, *Tajikistan*, http://www.worldbank.org/en/country/tajikistan (accessed 16-1-13).

65 Interfax News Agency, *RUSAL to help build aluminium smelter in Tajikistan*, 21 October 2004.

66 Interfax News Agency, *$ 1.5 bln into Tajikistan*, 1 February 2005.

67 Tajikistan, Russian investor seek World Bank expertise to solve dispute over power project, *Associated Press*, 19 April 2006.

68 Eurasianet.org, *Does Dushanbe Want to Distance Itself From Russia?*, http://www.eurasianet.org/departments/insight/articles/eav090707aa.shtml (accessed 10-10-13).

69 The World Bank, *Assessment Studies for Proposed Rogun Regional Water Reservoir and Hydropower Project in Tajikistan*, http://web.worldbank.org/WBSITE/EXTERNAL/COUNTRIES/ECAEXT/0,,contentM311DK:22743325~pagePK:146736~piPK:146830~theSitePK:258599,00.html (accessed 5-6-13).

70 Originally, the results of the study should have been released in the Summer of 2012.

71 The World Bank, *Assessment* (fn. 69).

identity and bolster patriotism, international criticism to the project and the necessity to mobilize financial resources. This rhetoric of justification, aimed at normalizing Rogun and at getting consent, will be discussed in the next section.

4. THE PEOPLE'S DAM

The Rogun ideology has been created and disseminated by Tajik President Emomali Rahmon and government officials from his close network of power, most notably the Foreign Minister Hamrokhon Zarifi, the Prime Minister Akil Akilov and the Tajik Permanent Representative to the United Nations (UN), Sirodjidin Aslov.[72] All of them have managed to keep an unvaried position towards the Rogun dam over the last decade, one that can be summarized into the motto "Rogun shall be built at all costs". Molle et al. noted that the creation of certain meta-discourses and meta-justifications – which usually tend to stress matters such as the achievement of national goals and priorities or the absence of real alternatives – are among the classical means of furthering large-scale projects.[73] This seems to apply also to the Rogun dam, which is typically presented by the Tajik political elite as a vital achievement, a key to energy-independence and generally as a solution to most of the problems faced by the country.

Thus, following the harsh energy crisis that hit Tajikistan in 2007/08, Emomali Rahmon started to recurrently discuss the Rogun dam in his addresses to the nation. This seems relevant to the propagation of the Rogun ideology, since the President's speeches are regularly disseminated and broadcasted by Tajik media such as the Khovar and Avesta news agencies, and the Tajik state-run TV and Radio. For instance in 2008, during a message to the Parliament, Rahmon launched an appeal:

> To every decent citizen, to all those noble people loving their homeland and nation, and especially – to experienced builders and specialists in the field of energy and construction, to take an active part in the speedy construction and commissioning of Rogun and contribute to energy independence, [...] a matter of life or death for the Tajik state.[74]

Furthermore, Rahmon pointed out that according to Islam and its sacred text the Qur'an,[75] water is at the origin of all life on Earth, and thus it has to be treated as a God-given gift for which humans should be grateful and respectful.[76] Therefore, while Sharia has no official status in Tajikistan,[77] Rahmon is using Islam as a legitimation tool, integrating it in his state ideology

72 In December 2013 Aslov was appointed Foreign Minister, thus replacing Zarifi as head of the Tajik MFA.

73 Molle et al., *Hydraulic* (fn. 23).

74 Rahmon, Emomali, *Message from the President of Tajikistan Emomali Rahmon to the Parliament*, http://www.president.tj/ru/node/867 (accessed 14-8-13), Dushanbe, 15 April 2008.

75 Sunni Islam is the official religion in Tajikistan.

76 Rahmon, *Message* (fn. 74).

77 Cummings, Sally N., *Understanding Central Asia. Politics and contested transformations*, New York, Routledge, 2012.

to justify a project that in his view would guarantee the best use of the nation's natural resources. Along the same line, in 2010 Rahmon centred on Rogun his yearly Presidential address to the people of Tajikistan:

> I shall reiterate to all citizens of this sovereign state, regardless of nationality, language and religion, that Rogun is a real battleground for honour and dignity, is a popular arena of selfless work for a better future and prosperity of sovereign Tajikistan! [...] I appeal to the children of Tajikistan, living and working in other countries, and always thinking about the welfare of their ancestral land and the prosperity of their houses: you can actively participate in this nation-wide initiative and contribute to the construction of Rogun, a source of light and heat in your homes! [...] a source of national pride for every citizen of Tajikistan and a symbol of pride for our present and future life! [...] a symbol of life and death of the Tajik state![78]

On this occasion Rahmon reiterated the appeal to his citizens, while also portraying the dam as a unifying element, a symbol of national pride and honour, a source of progress and prosperity, and eventually, as a matter of life and death for Tajikistan.[79] These elements of the discourse, and in particular the representation of the dam as an existential matter, have been used by Rahmon to persuade his citizens to purchase shares of the "Open Joint Stock Company Rogun"[80], that launched an initial public offering (IPO) on 6 January 2010, a day declared the "Day of Solidarity for the Construction of Rogun"[81]. During the IPO, Tajik citizens were asked to sacrifice part of their salaries to purchase shares of Rogun, while the main streets of Dushanbe had been adorned with banners and posters advertising the dam, and the Tajik state TV devoted substantial amounts of prime time broadcast to updates on the progress of the share sale.[82] The Tajik President repeatedly urged his executives to raise awareness about the sale, since this is the task of every official and of every honourable citizen.[83] Unsurprisingly, the Rogun IPO was severely criticised by the Uzbek government,[84] resulting in an epistolary dispute between the Uzbek Prime Minister Shavkat Mirziyoyev and his Tajik colleague Akil Akilov. Using a patriotic and nationalis-

78 Rahmon, Emomali, *Obrashcheniye Prezidenta Respubliki Tadzhikistan k Narodu Tadzhikistana*, http://khovar.tj/rus/archive/17084-obraschenie-prezidenta-respubliki-tadzhikistan-k-narodu-tadzhikistana.html (accessed 5-7-13), 5 January 2010.

79 Also, and in line with Arundaty Roy's analysis of the Sardar Sarovar Dam in India, this implies that if you support the dam you are a patriot, but if you don't, you are an enemy of the nation. Indeed, Suhrob Sharipov, the head of the Strategic Research Centre (SRC) of Tajikistan, was quoted as saying that "if somebody in the country opposes construction of the Rogun hydroelectric power station, he will automatically turn into a traitor" (BBC Monitoring Central Asia Unit, *Tajikistan, Uzbekistan should seek compromise on water row - Uzbek diplomat*, 16 June 2009).

80 Ministry of Finance of Tajikistan, *Obsuzhdeniya: Ob aktsiyakh i Sertifikatakh aktsiy OAO "Rogunskaya GES"*, http://minfin.tj/index.php?newsid=139 (accessed 3-3-13).

81 Eurasianet.org, *Tajikistan: Forced Rogun Payments Sowing Discontent Among Impoverished Tajiks*, http://www.eurasianet.org/departments/business/articles/eav010510.shtml (accessed-14-11-12).

82 One year later, some two million shares of Rogun had been sold, earning the GoT US$ 170 million, corresponding to less than 10% of the total amount required to build the dam.

83 Tajik leader urges officials to mobilize people to buy power plant shares, *BBC Monitoring Central Asia Unit*, 20 Janury 2010.

84 Ministry of Foreign Affairs of Uzbekistan, *Uzbek Prime Minister Writes to his Tajik Colleague on Rogun Hydrolelectric Power Station*, 4 February 2010.

tic rhetoric, Akilov stressed how the Uzbek letter united the "people of Tajikistan in the idea of building this vitally important hydropower plant"[85].

While the local media are the main way to spread the Rogun ideology at the national level, the channels to advocate for Rogun seem to be international forums like the UN and the European Union (EU), as well as conferences and seminars.[86] Starting in 1999 at the 54th United Nations General Assembly (UNGA), and carrying on uninterruptedly until the 68th UNGA in 2013, the GoT introduced water issues and subsequently Rogun in its annual address at the UN, in a strategy aimed at portraying itself as a responsible water user and as a global leader in encouraging cooperation in the field of water.[87] Hence, in the period 1999-2004, the GoT called for the attention of the world community on fresh water problems and on greater cooperation between countries, successfully putting forward two initiatives to declare 2003 the "International Year of Freshwater", and 2005-2015 the "International Decade for Action Water for life"[88]. The connection between Rogun and the achievement of the MDGs, which implies the representation of the dam as a fundamental element to attain national goals and priorities, became the central message delivered by Tajikistan at the UN during the subsequent years,[89] along with the launch of an initiative – the third of this kind in just a few years – to proclaim 2012 the "International Year of Water Diplomacy"[90]. Emblematically, at the most recent World Water Forum in France (2012), the GoT disseminated brochures and pens uttering the message "Tajikistan is a water country", thus consolidating the process aimed at binding the idea of Tajikistan with that of water.

5. CONCLUSIONS

This article has discussed how a "white elephant" (i.e., a large architectural project whose cost outweigh its benefits[91]) can be framed by a ruling elite in such a way that the project comes to symbolize national patriotism, co-

85 Tajikistan: Akil Akilov responded to official letter of Uzbek Prime-Minister on the Rogun project, *Ferghana News*, 8 February 2010.

86 Over the last decade, the GoT has organized several UN-backed water conferences in which the development of Tajikistan's hydroelectric potential featured a prominent role. Among them there are the "International Water Forum" in 2003, the "International conference on regional cooperation in transboundary river basins" in 2005, the "Water for Life" conference in 2010, and the conference "Towards the conference on sustainable development (RIO+20): water cooperation issues" in 2011.

87 Rahmon, Emomali, Statement by the President of the Republic of Tajikistan H.E. Mr. Emomali Rahmon at the 63rd Session of the UN General Assembly, New York, 25 September 2008.

88 Rahmon, Emomali, Statement by the President of the Republic of Tajikistan H.E. Mr. Emomali Rahmon at the 54th Session of the UN General Assembly, New York, 1 October 1999; Rahmon, Emomali, Statement by the President of the Republic of Tajikistan H.E. Mr. Emomali Rahmon at the 58th Session of the UN General Assembly, New York, 30 September 2003; Nazarov, Talbak, Statement by the Minister of Foreign Affairs of the Republic of Tajikistan Academician Talbak Nazarov at the 59th Session of the UN General Assembly, New York, 30 September 2004.

89 See for instance Rahmon, Emomali, Statement by the President of the Republic of Tajikistan H.E. Mr. Emomali Rahmon at the 64th Session of the UN General Assembly, New York, 23 September 2009.

90 Rahmon, Emomali, Statement by the President of the Republic of Tajikistan H.E. Mr. Emomali Rahmon at the 64th Session of the UN General Assembly, New York, 23 September 2010.

91 For an overview of white elephants see Robinson, James A./Torvik, Ragnar, "White elephants", in: *Journal of Public Economics*, vol. 89, 2005, 197-210.

hesion and identity. The notions of progress and modernity, along with the assertion of the nation's right for self-determination, are being extensively used by the Tajik government to persuade its citizens that the Rogun dam can indeed be the panacea for cash-strapped Tajikistan. Once again, the thaumaturgic properties ascribed to a large dam are part of a discursive process that tends to overlook the numerous downsides that can come along these kinds of projects. If on the one hand the Rogun dam could almost double Tajikistan's energy production, on the other hand a similar result could be obtained with more economical and less controversial solutions such as small hydropower stations. These solutions, however, are considerably less spectacular and politically appealing than the tallest dam in the world.

Yet, megaprojects – and more in general grandiose developments plans such as for instance the ones carried out by the Soviet Union – can sometimes be realised out of real necessity, coming to represent a paramount achievement for those who built them. The Øresund Bridge between Sweden and Denmark, for example, not only proved economically viable experiencing substantially more traffic than was expected in its first years, but also symbolically connected Sweden to continental Europe.[92] In other cases, the megaproject can still embody a strong symbolic value while nevertheless proving to be fundamentally uneconomical, as the Channel Tunnel linking Britain to France. Also, a megaproject can be both unnecessary and uneconomical, as the proposed Strait of Messina bridge in Italy and, presumably, the Arena da Amazônia stadium that was built in Brazil for the 2014 Football World Cup.

The case of the Rogun dam goes across most of the abovementioned instances. The dam could indeed prove to be an excellent solution for Tajikistan's energy problems just as much as it could result in a geopolitical and economical misadventure. The discursive construction put in place by the Tajik government to portray it as a national symbol remains, at present, the most concrete element in the history of this project.

LIST OF FIGURES

92 Flyvbjerg, Bent/Bruzelius, Nils/Rothengatter, Werner, *Megaprojects and Risk. An Anatomy of Ambition*, Cambridge 2003.

PART 4: COLONIES UNDER CONSTRUCTION

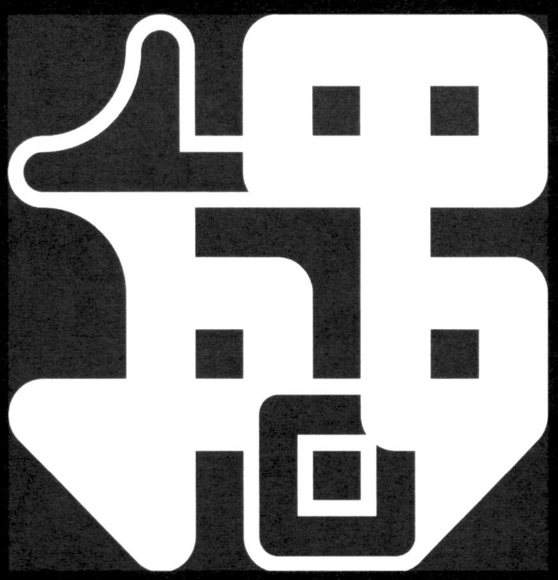

RESHAPING THE VENETIAN TERRA FERMA

REMARKS ON THE LITERARY, PICTORIAL AND ARCHITECTURAL PERCEPTION OF EARLY MODERN LANDSCAPE TRANSFORMATION IN ITALY

Sören Fischer

1. INTRODUCTION

Venice's hinterland, the Terra ferma, was an important construction site in Early Modern Europe, both in its sheer scope as in its widely distributed representation. By the use of artificial canals, dikes, machines and new roads no other Italian region was in the sixteenth century altered in its shape, structure and natural character in such intensity.[1] Wide areas of the Venetian home land became construction sites that not only changed the material world radically but were also perceived as objects of an elaborate discursive and performative strategy that put the change of landscape, economy and society on display. Analysing the ways construction sites and technological changes were connected with the elements of advertising, propaganda and idealization in sixteenth century Venetian land cultivation proofs a paradigmatic role model for the centuries to come.

The efforts of cultivating the mainland became necessary when the Ottoman offences against the state's foreign territories and trading posts such as the island Rhodes (conquered by the Ottomans in 1523) in the Mediterranean Sea as well as the shifting of the international trade routes to the Atlantic region compromised the Republic's stability.[2] The state of Venice and its population had been dependent on shipping to guarantee the permanent provision of corn, maize and rice – especially in war times. Built on many islands, the capital itself lacked large areas for agriculture, and its continental soil – the Terra ferma – was due to mosquitoes and diseases in many regions insalubrious marshland that could not feed the Venetian population on its own.

As Denis Cosgrove and Salvatore Ciriacono have shown, one of the most intensive and complex enterprises in land cultivation and fertilization in Early Modern Europe took place – mainly promoted by the author and land owner Alvise Cornaro, who belonged next to Gian Giorgio Trissino

1 For introduction to the policy of land reclamation in sixteenth century Venice see Fontana,Vincenzo, "Alvise Cornaro e la terra", in: Puppi, Lionello (ed.), *Alvise Cornaro e il suo tempo, exhibition catalogue*, Padova 1980, 120-128; Ciriacono, Salvatore, "Irrigazione e produttività agraria nella Terraferma veneta tra cinque e seicento", in: Archivio Veneto, CXII, 1979, 73-135; Ciriacono, Salvatore, *Acque e agricoltura nell'Europa moderna. Il caso veneziano*, Padova 1992; Cosgrove, Denis, *Il paesaggio palladiano*, Verona 2004, 238; Ciriacono, Salvatore, *Building on Water. Venice, Holland and the Construction of the European Landscape in Early Modern Times*, New York/Oxford 2006; Moriani, Gianni, *Palladio architetto della villa fattoria*, Verona 2008, 211-244; Smienk, Gerrit/Niemeijer, Johannes (ed.), *Palladio, the Villa and the Landscape*, Basel 2011.

2 Moriani, *Palladio architetto* (fn. 1), 13-20; Fischer, Sören, "The Allegorical Landscape: Alvise Cornaro and his Self-Promotion by the Landscape Paintings in the Odeo Cornaro in Padua", in: *Kunstgeschichte. Open Peer Reviewed Journal* 2013, 1-2.

and Daniele Barbaro to the high class of humanistic society in the Venetian Republic.[3] This transformation, inspired also by the land cultivation that started in Holland in the early sixteenth century, primarily was overseen and administrated by the so called *Magistratura sopra i Beni Inculti*, an organisation that was founded in 1556 by the Venetian senate.[4] For the *Serenissima* this meant an elementary shift in its policy. Venice had been a republic primarily orientated towards the Sea up to this point.

It was the Venetian nobility that conducted these efforts in the first place. But also various noble and rich families from the Terra ferma's elite itself – such as from Padua, Vicenza, Treviso or Rovigo – followed the new idea of agriculture known as "Santa Agricoltura". In consequence, from 1530 on wide areas of the Terra ferma became construction sites that completely changed the landscape's morphological, geological as well as its cultural appearance.

By analyzing the multilayered impact of these interventions, the present essay highlights the relationship between the physical transformation of the Venetian landscape in the sixteenth century and its perception in cartography, literature, architecture and painting. The central question is: In what ways and in what intensity were the efforts of reshaping nature perceived? And already at this point it is to say that we are dealing with a complex two-layer perception that differentiated between the real material world and its interpretation in the arts.

2. RESHAPING THE TERRA FERMA – MAPS, MACHINES, MANPOWER

The extensive cultivation of the Terra ferma is exemplified by the activities in the area of Polesine close to Rovigo. Between 1533 and 1541, nearly 90.000 *campi padovani* (c. 34.200 acres) were cultivated by canalization and water mills.[5] In only a few decades rare habitats for animals and plants were altered, partly destroyed and replaced by large-scale monocultures of rice, corn and grain. They allowed the landowners a highly intensified agriculture and an increase in harvests. It seems symptomatic for the positivistic approach towards that policy that we have no evidence of a critical discussion of this environmental alteration and change of the status quo though.

These efforts were accompanied by cartography. The detailed map by Cristoforo Sorte, drawn in 1556 in a bird's eye view, documents the agricultural landscape and the many rivers, canals and fields in the northern part of the Republic [fig. 1].[6] The entire land is organized by these artificial structures. It almost appears as if the progress of land cultivation has written its own new texture on the earth's surface. Thus, the entire countryside became subject to change.

3 Introducing to the biography of Alvise Cornaro see Fiocco, Giuseppe/Cornaro, Alvise, *Il suo tempo e le sue opere*, Vicenza 1965; Blason Mirella, "La vita di Alvise Cornaro", in; Puppi, Lionello (ed.), *Alvise Cornaro e il suo tempo, exhibition catalogue*, Padova 1980, 18-26; Lippi, Emilio, *Cornariana. Studi su Alvise Cornaro*, Padova 1983.

4 Ciriacono, *Building on Water* (fn. 1), 1-18.

5 See Moriani, *Palladio architetto* (fn. 1), 30.

6 See Smienk/Niemeijer, *Palladio* (fn. 1), 16. The cartography as part of the transformation process in the Terra ferma was analysed by Cosgrove, *Il paesaggio palladiano* (fn. 1), 245-273.

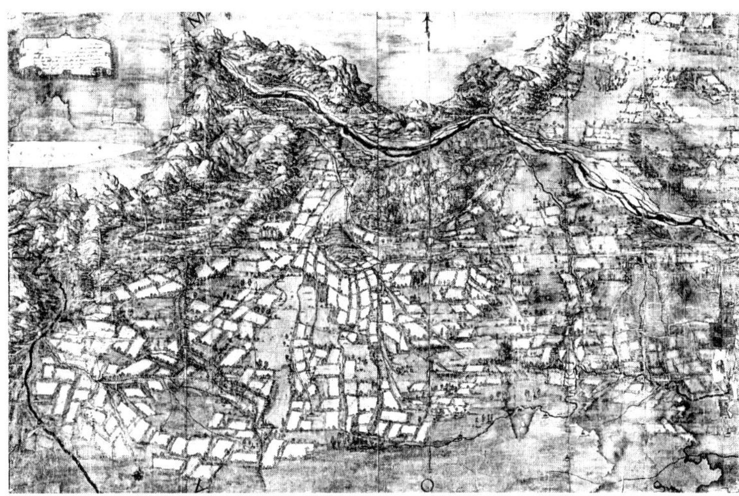

Fig. 1 Cristoforo Sorte, Map of the north region of the Veneto, 1556.

The impacts on nature and on civil life during that time were fundamental. In only a few decades, 500.000 *campi* (c. 190.000 acres) around Padua, Treviso, Rovigo and Aquileia were cultivated.[7] 250 new villages and many splendid villas arose on a land that recently had been uninhabitable, in many regions even dangerous and inhospitable due to diseases. The newly created farmland also had an effect on the population. In 1548, the number of inhabitants of the Terra ferma (the city Venice excluded) was 1.417.000. In 1565, it had already risen to 1.500.000, and reached the number of 1.573.000 in the year 1625.[8]

The intervention into the landscape was particularly strong in the area between the Colli Berici and the Colli Euganei, and was documented in a map by Antonio dall'Abaco in 1567 [fig. 2].[9] In only a few decades a new

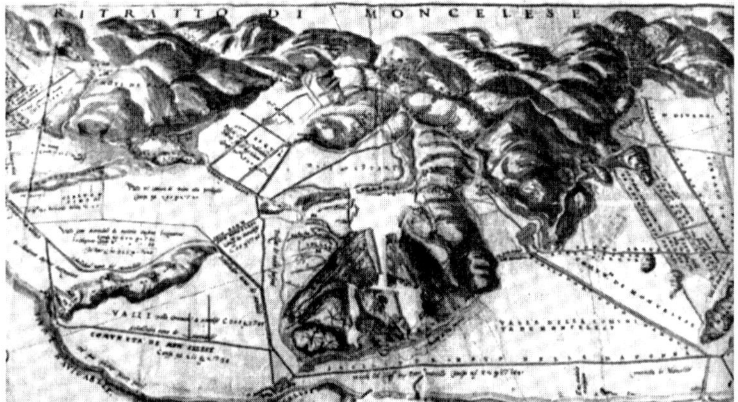

Fig. 2 Antonio dall'Abaco, Map of the area of Monselice, 1567.

7 Fontana, *Alvise Cornaro* (fn. 1), 120-128.

8 Cosgrove, *Il paesaggio palladiano* (fn. 1), 207, n. 1.

9 Smienk/Niemeijer, *Palladio* (fn. 1), 7-25.

and complex infrastructure of waterways and small canals was created which changed the economic viability of the region. These new watercourses allowed not only the systematically drainage and reclamation of low land, they also served for efficiently transporting agricultural products.

Also Andrea Palladio (1508-1580), the most influencing architect of the Venetian villa architecture, underlined the usefulness of waterways in his architectural treatise *I Quattro Libri dell' Architettura*, published in Venice in 1570. Picturing the perfect location for a modern country house he wrote in the chapter *Del sito da eleggersi per le fabriche di villa* ("On the place that should be selected for country houses"):

> It will be most convenient and attractive if it [the villa] can be built on a river, because the produce can be carried cheaply by boat to the city at any time, and it will satisfy the needs of the household and the animals; this will also make it very cool in the summer and will be a lovely sight, and is both useful and pleasing in that one can irrigate the grounds, the gardens, and the orchards [bruoli], which are the soul and delight of the estate."[10]

An illustration drawn and published by Geritt Smienk and Johannes Niemeijer in their book on the Palladian landscape in the sixteenth century elucidates the character of this process [fig. 3].[11] The parcellation pattern clearly shows that the reshaping of the Terra ferma was not a result of a general plan, but of various private initiatives, conducted by the many rich Venetian families. Around 1560 Alvise Cornaro praised the miraculous effect of the reclamation on the Venetian culture and on the population:

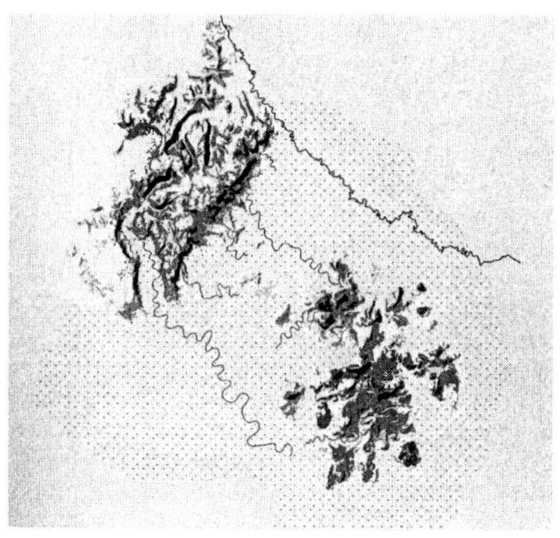
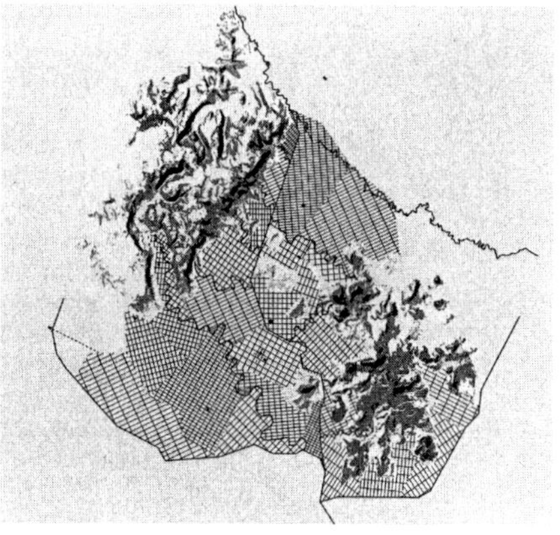

Fig. 3 Schematic illustration of the reclamation system between the Colli Berici and the Colli Euganei in the sixteenth century.

10 Palladio, Andrea, *I Quattro Libri dell'Architettura*, Venice 1570, II, 12, 45-46: "Se si potrà fabricare sopra il fiume, sarà cosa molto commoda, e bella: percioche e le entrate con poca spesa in ogni tempo si potranno nella Città condurre con le barche, e servirà a gli usi della casa, e de gli animali, oltra che apporterà molto fresco la Estate, e sarà bellissima vista, e con grandissima utilità, & ornamento si potranno adaquare le possessioni, i Giardini, e i Bruoli, che sono l'anima, e diporto della Villa." The translation is cited by Palladio, Andrea, *The Four Books on Architecture*, transl. by Robert Tavernor and Richard Schofield. Cambridge 2002, II, 12.

11 Smienk/Niemeijer, *Palladio* (fn. 1), 20-21.

And in less than two years the healthy air had returned. And the 40 souls that had lived in this region before the reclamation are now up to 2000. Indeed I can say that I gave God an altar and a temple and many souls to worship him.[12]

Cornaro, who used a religious terminology, stylized himself as a creator who gave the formerly poor population housing and food. However, it is symptomatic of the public representation of the Venetian land cultivation that the work and living reality of the rural population did not play any decisive role. Instead, cultivation is represented as an act of godlike creation. The strong and programmatic references to the Genesis is obvious also in a note of the *Magistratura sopra i Beni Inculti*. This document was written as a preparation for land cultivation and discusses the different steps of that process around Monselice. The terraforming of the land appears as an enterprise that not only was supported by the Venetian nobility and the entire state, but by God itself:

The cultivation of land has to follow the example of God's creation in three steps. When he created the world he primarily divided the sky from the other matter, than the ground from the water. Finally he created various beings: Animals, trees and grain. Any cultivation of land has to follow his plan in three steps. First the water has to be diverted out of the ground [...].[13]

A map with the water system that was drawn by Gianbattista Dante around 1540 impressively illustrates the high efforts and complex technical activities [fig. 4].[14] During the campaigns it was not only necessary to dig canals but also to invent a system that allowed diverting and separating water on

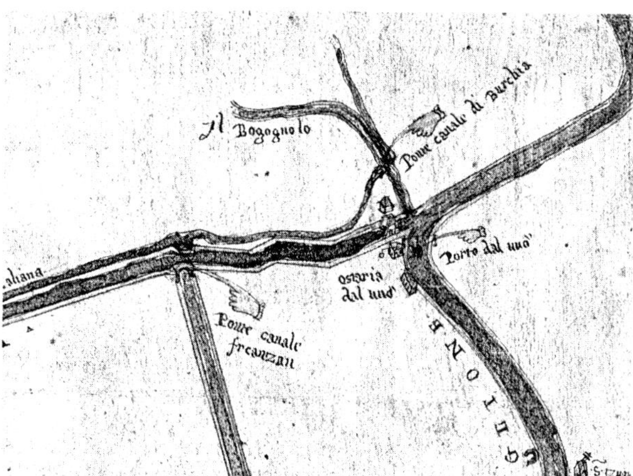

Fig. 4 Gianbattista Dante, Map with water system between the Colli Berici and the Colli Euganei, circa 1540.

12 Cornaro, Giacomo Alvise, *Elogio*, cited after Fiocco, Giuseppe, *Alvise Cornaro. Il suo tempo e le sue opere*, Vicenza 1965, 200-203, here 200: "[...] et in men di due anni li reduse tuti al coltura, et ritornò il buono aere a quelle villa e luogo nel quale tanto era lo aere tristo che non si potea conservare ritti li pini che naseano. Ma levate le aque il male aere cessò et vene il buono. Et di 40 anime che vi erano, ora ve sono due milia [...]. [...] in tal luogo desse a Dio altare et anime per adorarlo [...]."

13 Cited after Muraro, Michelangelo/Morton, Paolo, *Villen in Venetien*, Köln 1999, 43.

14 Smienk/Niemeijer, *Palladio* (fn. 1), 20.

two different levels. Machines capable to lift, transport and control water were of great importance in this process.[15] Some such machines that were powered by horse and manpower were displayed by Daniele Barbaro and Giuseppe Ceredi between 1556 and 1567 [fig. 5].[16] In Giuseppe Ceredi's illustration the positive impact of the water regulation is particularly expressed by the various plants in the gardens and fields. They symbolically represent fertility and prosperity.[17]

Fig. 5 Giuseppe Ceredi, Machine to transport water, powered by a horse, in: Tre discorsi sopra il modo d'alzar acque da' luoghi bassi, Parma 1567.

15 As also described by Ciriacono, *Building on Water* (fn. 1), 7: "[...] Venice was lacking an agriculture that fully exploited its water resources to boost fodder production, increase herd size and thus achieve higher agricultural yields".

16 Barbaro, Daniele, "Machines to elevate and transport water", in: *I dieci libri dell'architettura di M. Vitruvio*, Venice 1556, 265; Ceredi, Giuseppe, "Machine to transport water, powered by a horse", in: *Tre discorsi sopra il modo d'alzar acque da' luoghi bassi*, Parma 1567, 74-6; see also Ramelli, Agostino, "Machine to lift water, powered by a stream", in: *Diverse et artificiose machine*, Paris 1588.

17 Ceredi, "Machine to transport water" (fn. 16), 74-76.

3. THE VILLA AS AN ARCHITECTURAL LANDMARK

Next to these large scale technical operations, many Venetian villas created interventions in the landscape, as also recently elaborated by Smienk and Niemeijer. With these administrative centre's substructures, earthworks, rerouting of streets and streams, with their monumental architecture and domination of the country side, the buildings changed the perception of the Venetian Terra ferma and became new landmarks. The villas' architectures were staged as the most programmatic expression for the drastic rearranging and forming of the material world, as will be illustrated by the Villa Pisani in Bagnolo, the Villa Emo in Fanzolo and the Villa Barbaro in Maser, all constructed by Andrea Palladio.

When Palladio built the Pisani Villa somewhere between 1542 and 1544 he placed it in the centre of the family's farmland, allowing for the estate's easy administration and visual supervision. As a map from 1569 illustrates, the reclaimed fields were situated on both sides of the river with a total area of 1.200 *campi* (500 hectares).[18] Already in this early project Palladio aimed to connect the local infrastructure with the Villa. It is not only placed along the navigable river, the Villa is also situated in the intersection of the three main streets.[19] Placed like this, it is until today impossible for any traveller to miss the villa as the landscape's central protagonist. The villa shapes the landscape in the flat plane of the Terra ferma – a landscape of noble architecture *all'antica*.

A similarly strong architectural intervention is the Villa Emo, completed around 1565 by Palladio for Leonardo Emo and his family. The building confronts the flat countryside with a nearly 100 meter long impressive façade. With its representative size and the creation of dominant visual axes the building is, as studied also by Smienk and Niemeijer, literary inscribed into the region's parcellation pattern.[20] The long alley works as a natural prolongation of the Villa and reflects the cultural value of the man-made architecture into the landscape, underlining the concept of a transformed and cultivated world.

A comparable approach towards agricultural landscape and architecture we observe with Villa Barbaro. Build for Marcantonio and Daniele Barbaro from circa 1557 to 1560 it intentionally altered the landscape and its perception. Artificial itself it becomes – as the Villa Pisani and the Villa Emo – the very signature for a technically transformed nature. Only at first sight the Villa seems to be organically attached to the hill behind it. Yet a closer look reveals that the architect performed strong interventions into the topography. Palladio not only structured the terrain in different levels and established a long axial connection between the country house and the wide plane of Maser, he also created an elaborate water system.

The programmatic and ideologically charged connection of landscape and villa becomes more obvious when we analyse the vistas as seen from

18 Smienk/Niemeijer, *Palladio* (fn. 1), 51.

19 Ibid., 55.

20 Ibid., 107.

inside many Venetian villas.[21] Standing in the central room of Villa Emo the observer can't see the wide panorama that is visible from the outside [fig. 6]. Instead, the landscape gets segmented into three rectangular windows that are framing the farmland like pictures would. The agricultural land shifts into the appearance of an image, of a painting. The material reality becomes an artificial quality. Villa Barbaro's design uses the same effect. Here, the windows limit the cultivated countryside and convert it to nearly pictorial illustrations of agriculture: The cultivated landscape is perceived as a painting.[22]

The transformation of the Terra ferma and the systematic reshaping of its surface happened through the reclamation of land, in the drawing of detailed maps and in the use machines. However, primarily the Venetian villa was staged as an architectural landmark and as an expression for the change of the material world. Its architecture and visual dialogs to the environment have to be interpreted as statements on the contemporary reclamation efforts. The Villa was directed as a place of a strong ideological and programmatic concretion, its presence became the manifestation of the impact, technology and ideology had on the landscape's shape.

Fig. 6 Andrea Palladio, Villa Emo, Fanzolo, c. 1565: Vista from the interior on the main alley and the fields.

21 For the aesthetics of the framed vista in the Venetian villa see also Fischer, Sören, "Fingere alcune aperture, in quelle far paesi da presto e di lontano" – On the Concept of the Framed Vista in the Early Modern Italian Villa", in: Kacunko, Slavko/Harlizius-Klück, Ellen/Körner, Hans (ed.), *Framings*, Berlin 2015 (in preparation for print).

22 For the topos of the window as a painting in the Venetian villas see Fischer, Sören, *Das Landschaftsbild als gerahmter Ausblick in den venezianischen Villen des 16. Jahrhunderts. Sustris, Padovano, Veronese, Palladio und die illusionistische Landschaftsmalerei*, Petersberg 2014, 78-129; see also Blum, Gerd, "Palladios 'Villa Rotonda' und die Tradition des 'idealen Ortes'. Literarische Topoi und die landschaftliche Topographie von Villen der italienischen Renaissance", in: *Zeitschrift für Kunstgeschichte*, vol. 70, 2007, 159–200; Blum, Gerd, "Fenestra prospectiva. Das Fenster als symbolische Form bei Leon Battista Alberti und im Herzogpalast von Urbino", in: Poeschke, Joachim/Syndikus, Candida (ed.), Leon *Battista Alberti. Humanist, Architekt, Kunsttheoretiker*, Münster 2008, 77–122.

4. STAGING A VISIONARY WORLD: THE ILLUSIONISTIC LANDSCAPE PAINTINGS AS COMMENTS ON THE RECLAMATION EFFORTS

The illusionistic landscape paintings that cover the walls of various Villas show, however, that the perception of the reclamation process was much more sophisticated. These frescoes – such as in the Villa dei Vescovi from around 1542-1543, the Villa Godi from 1548-1550 and in the Villa Barbaro from 1560-1561 – depict lovely countrysides, ancient ruins, green planes and peaceful shores, an overall flourishing and fertile nature.[23] The illusionistic architectures, which function as frames for the painted landscapes, support the impression that the observer sees through real windows.

As for example in the Sala dei Cesari in the Villa Godi, decorated by the local painter Gualtiero Padovano around 1549 for Girolamo Godi and its family, frescoes open the walls of the room on to various landscapes [fig. 7]. The entire room is mutating into a free standing loggia that allows a panoramic view of an imagined world. Framed by painted architecture the observer not only sees an idealized nature, but also a population who enjoy the peace and the tranquillity of the countryside [fig. 8]. One sees noble men and women taking a trip on a boat, a young man enjoying an ideal resting spot under a tree – a *Locus amoenus* – and a noble lady riding a horse. These frescoes impressively picture the idea of the so called *Piaceri della Villa* (the Leisure activities in the country side) that were guarantors

Fig. 7 Gualtiero Padovano, Sala dei Cesari, Villa Godi, Lugo di Vicenza, 1548/1550.

23 For the illusionistic landscape paintings in these villas see Fischer, *Das Landschaftsbild* (fn. 22), 92-127; Fischer, Sören, "Das Odeo Cornaro, die Villa Farnesina und der Oecus kyzikenos. Wandmalerei und Architektur im Spannungsfeld zwischen vitruvianischer Textexegese und Interpretation", in: *Zeitschrift für Kunstgeschichte*, vol. 77, 2014, 199-220.

Fig. 8 Gualtiero Padovano, Sala dei Cesari, Villa Godi, Lugo di Vicenza, 1548/1550, detail.

of an ideal villa life. In the villa books of the sixteenth century these *Piaceri* are among the most important *leitmotivs*. In the book *Le dieci giornate della vera agricoltura e de' piaceri della villa*, published in 1564, Agostino Gallo propagated the villa as a perfect living place:

> Who would not love to live in a villa, where you can find peace, true liberty, tranquility and a sweet place for resting? Here you enjoy also the fresh airs, the leaves of the trees, their fruits, the purity of the springs [...], the fertility of the acres [...] and the beauty of the gardens.[24]

Some years earlier, in 1558, also Alvise Cornaro had chosen similar words to describe his villa life in the Colli Euganei south to Padova:

24 Gallo, Agostino, *Le dieci giornate della vera agricoltura e de' piaceri della villa*, Venezia 1565, VIII, fol. 169r: "Chi non doverebbe adunque habitare in villa; poiche non tanto vi si trova la buona pace, la vera libertà, la sicura tranquillità, & ogni soave riposo; ma vi si gode anco l'aprico aere, le frondi de gli arbori, i fruttiloro peregrini, la chiarezza dell'acque, l'amenità delle valli, la prospettiva de' monti, l'allegria de' colli; la vaghezza de' boschi, la spatiosità delle campagne, la fertilità delle possisioni; la utilità delle viti, e la bellezza de' giardini."

Next to these things I have found another way to amuse me. In April and March as in September and October I enjoy for some days my hill that is located in the area of the Colli Euganei. And in the most beautiful spot of this area I have fountains, gardens and above all a very comfortable and nice room [...].[25]

In Gallo and Cornaro the cultivated agricultural land transforms to a Garden Eden, to a *Paradiso terrestre*, as it was put in words highly ideological in the sixteenth century.[26]

However, like on the frescoes in the Villa Godi, there was no room for the life of the peasants and their hard work on the fields in the literary representations of villa life and land cultivation.[27] The agricultural reality, the various efforts to transform the Venetian homeland and the everyday work of the farmers, as it was illustrated for example in Fra Giacondo's commentary on Vitruvius (published for the first time in 1511) or in a contemporary drawing, attributed to the circle of Domenico Campagnola, was totally neglected [fig. 9, 10].[28] In several Venetian villas, such as in the Villa dei Vescovi (1452/43) and the Villa Barbaro (1560/61) decorated by Lambert Sustris and Paolo Veronese, painted landscapes appear as bucolic spaces, inspired by Vergil' idealization of rustic life [fig. 11, 12].[29]

Wall decorations and literary works both underline the assumption that the process of land cultivation and the mundane job of the peasant population as the nucleus of villa life were not adequate subjects to be painted onto the walls or to be described in books. The technical interventions, the alterations of the genuine territory are categorically hidden in the interior

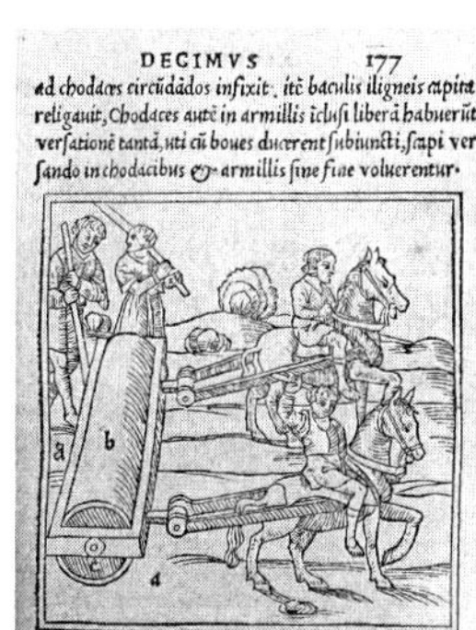

DECIMVS 177

ad chodaces circũdãdos infixit. itẽ baculis iligneis capita religauit, Chodaces autẽ in armillis ĩclusi liberã habuerũt versationẽ tantã, uti cũ boues ducerent subiuncti, scapi versando in chodacibus & armillis sine fine voluerentur.

Fig. 9: Illustration of agricultural work in: *Fra' Giacondo, Commentary of Vitruvius*, 1523 (left).
Fig. 10: Circle supposedly of Domenico Campagnola,Threshing with the flail, drawing, undated, c. 1540 (right).

25 Cornaro, Alvise, *Trattato de la vita sobria del magnifico M. Luigi Cornaro Nobile Vinitiano*, Venezia 1558, fol. 21r: "Ho anchora oltre à questo un'altro modo di solazzarmi, che io vò l'Aprile e'l Maggio, & così il Settembre, & l'Ottobre, per alquanti giorni à godere un mio colle, che è in questi monti Euganei, & nel più bel sito di quelli, che ha le sue fontane & giardini, & sopra tutto commoda et bella stanza [...]."

26 See Fischer, *Das Landschaftsbild* (fn. 22), 60-4.

27 See also Benzoni, Gino, *Gli affanni della cultura. Intellettuali e potere nell'Italia della Controriforma e Barocca*, Milan 1978, 134–143.

28 Vitruvius /Giacondo, Fra/Frontinus, *M. Vitruvii De architectura libri decem [...]*, Lyon 1523, fol. 177; Moriani, Palladio architetto (fn. 1), 113.

29 Introducing to the landscape paintings in these buildings see Fischer, *Das Landschaftsbild* (fn. 22), 110-127.

Fig. 11: Lambert Sustris, Stanza del Fanciullo, Villa dei Vescovi, Luvigliano, 1542/43.
Fig. 12: Paolo Veronese, Sala del'Olimpo, Villa Barbaro, Maser, 1560/61.

decorations. We see persons of the Venetian nobility enjoying a work-free day in a perfect nature, but no dykes, canals, machines field parcellations or any other kind of men-mad invention. The painted windows therefore intended to put landscapes on display that were deliberately different from the material world.

5. CONCLUSION

Finally this essay can summarise that the perception of the interventions into the physical landscape of the Terra ferma was divided in two separate levels. The first is constituted by the detailed, documentary cartography, the illustrations of machines, theoretical texts and the villa buildings. These medias are direct results of the change that took place in the material world. At the same time they are testimonies to the constant transformation and improvement of the rural landscape. Thus they document a future-oriented process that served not only the welfare of the landowners and their own interests of making money, as formulated very unbalanced for example by Bentmann and Müller in their influential study "Die Villa als Herrschaftsarchitektur. Versuch einer kunst- und sozialgeschichtlichen Analyse" in 1970.[30] Creating a reliable supply of the population with food this process acted also as a stabilizing moment that consolidated in a macroeconomic way the entire state and functioned as one important foundation of the Venetian political stability in the decades and centuries to come. Committed to that ideology the illusionary frescoes do not focus on the production of an economically productive countryside, but depict a status of completion. They do not portray reality but picture a visionary, nearly static world, an imaginary landscape. What the observers saw when they entered the Venetian villas were vistas into the Terra ferma's utopian future. The motive of constructing a landscape was transformed into the vision of a flourishing nature, a harmonious peaceful *Vita in villa* and the rekindling of the idea of the antique country life becoming reality only by means of a successful policy of land cultivation. Such land cultivation would lead, as predicted by Cornaro, to a world that reveals its original beauty. The illusionistic landscape paintings are therefore a vital element to understand the contemporary interpretation, idealization and the propagandist quality of the reshaping of the Venetian Republic.

The Venetian process of land cultivation – that means the Terra ferma as a material and mental construction side – has the character of a model, which is transferable to subsequent centuries. The consequences of technological changes were highly idealized by the state and the Venetian noble families, the most influential opinion makers of the time and the region. In particular, the burdens of the ordinary people have been ignored in the public perception and promotion, and veiled were also the vital interests of land speculation and money-making.[31]

30 Bentmann, Reinhard/Müller, Michael, *Die Villa als Herrschaftsarchitektur. Versuch einer kunst- und sozialgeschichtlichen Analyse*, Frankfurt a. M. 1970.

31 See also the study of Bentmann/Michael Müller, *Die Villa als Herrschaftsarchitektur* (fn. 30), 22-28.

Both Alvise Cornaro and the *Magistratura sopra i Beni Inculti* remain silent about negative effects on the rural population, about conflicts and possible resistances towards the technological, environmental and social changes. In this way they are significant examples of a positivistic, proto-colonial legitimization for land acquisition, land ownership, cultural replacement and domination. The Terra ferma as a building side programmatically produced, established and consolidated various pictures and ideas that helped to implement and to justify a new order, a new power and a new social hierarchy on the Venetian mainland.

LIST OF FIGURES

PART 5: REPRESENTATION UNDER CONSTRUCTION

CONSTRUCTING A DISCOURSE ON THE ART OF ENGINEERING: DOMENICO FONTANA AND THE VATICAN OBELISK

Clemens Voigts

One of the most famous depictions of a construction site shows the transport and the erection of the Vatican obelisk in St. Peter's Square in 1586 [fig. 1]. With this project Domenico Fontana (1543-1607) entered the annals of engineering by constructing the required lifting devices and conducting the entire operation. Up to the present day the erection of the Vatican obelisk is considered to be a milestone in construction history.

This article will investigate the circumstances that led to the outstanding significance of the project. It can be shown that Domenico Fontana's reputation is not only the consequence of a remarkable technical achievement *per se*, but also is a result of how he publicised it. In this regard, the book that he published in 1590 played a pivotal role: it provided a detailed report on the project and depicted every stage of it in elaborate etchings.[1] In the following, not only is Fontana's technical *modus operandi* re-examined, but it is also shown how his personal actions became a point of reference

Fig. 1 Construction site of the Vatican obelisk on St. Peter's Square (Carlo Fontana, 1694).

1 Fontana, Domenico, *Della trasportatione dell'obelisco vaticano et delle fabbriche di Nostro Signore Papa Sisto V fatte dal Cavallier Domenico Fontana architetto di Sua Santità*, Rome 1590.

for engineers of later generations. An analysis will be given of subsequent projects for the transportation and lifting of heavy loads: the Sigismund-Column in Warsaw as well as the famous pediment stones of the Louvre, the so-called Thunder Stone of Saint Petersburg as well as the obelisks of Paris, London, and New York. These examples will demonstrate how with both the technical solution for the moving of the Vatican obelisk and its public presentation Domenico Fontana began a discourse on the art of engineering that continued well into the nineteenth century.

1. A LEGENDARY INVENTION

In his report on the moving of the obelisk Domenico Fontana first of all gives us an account of how he was commissioned with the project. On 24 August 1585 Pope Sixtus V (1521-1590) formed a committee that should determine the best method for the transportation of the *guglia*, as the obelisk was commonly called. On the same day the committee decided to advertise a competition in order to gather the best ideas of "all available scholars, mathematicians, architects, engineers, and other qualified men"[2]. On 18 September, as Fontana tells us, his project was chosen by the jury from about 500 competing proposals. Unlike most of his competitors, who wanted to transport the *guglia* in an upright position, Fontana planned to lift the obelisk from its old position south of St. Peter's by means of a huge scaffold and to lay it down onto a sliding bed on which it would be transported horizontally. Finally, the obelisk should be re-erected in St. Peter's Square with the help of the scaffold.

Fontana's success in the competition might have seemed apparent, since for many years he had been the personal architect of Cardinal Felice Peretti Montalto who in April 1585 was elected as Pope Sixtus V. However, modern historical research has brought to light that Fontana whitewashed the facts with his anecdote of the competition. As Cesare D'Onofrio pointed out, Fontana needed more than one attempt to elbow out his competitors, in particular the famous Florentine sculptor and architect Bartolomeo Ammannati (1511-1592).[3] In fact, Ammannati had been entrusted with the moving of the obelisk already by 16 August 1585, even before the competition was announced. Also on 18 September, when the selection committee came to its final decision, the majority was again in favour of Ammannati's project. Only after Fontana had called on his papal patron did Sixtus V personally intervene and commissioned Fontana with the project on 25 September.

For Ammannati this was in fact his third defeat by Fontana within only a few months: in the summer of 1585 Domenico had snatched away from him the contracts for the *Cappella Sistina* in *Santa Maria Maggiore* and for the Lateran Palace.[4] However, in the case of transporting the obelisk,

2 Fontana, *Della trasportatione* (fn. 1), fol. 5 r: "tutti li litterati, Mattematici, Architetti, Ingegnieri, & altri valent'huomini, che si potessero hauere".

3 D'Onofrio, Cesare, *Gli obelischi di Roma*, Rome [2]1967, 72-77 for a detailed reconstruction of the events and for references.

4 Fagiolo, Marcello, "Ammannati e Sisto V: L'Obelisco Vaticano, la Cappella Sistina, il Palazzo Lateranense", in: Rosselli Del Turco, Niccolo (ed.), *Bartolomeo Ammannati: Scultore e architetto: 1511-1592*, Florence 1995, 195-207.

Ammannati was much more qualified since he was the only architect who could rightfully claim to have some experience in moving and raising large monoliths. Under his direction a granite column from the Baths of Caracalla in Rome, which Pope Pius IV had given the Medici family as a present, was transported to Florence and erected in 1565.[5] Of course, this column was not even half as tall as the *guglia* and weighed only about a quarter of it; so in the final commissioning Ammannati's expertise was outweighed by Fontana's close relationship with the pope.

In this way several other highly respected and influential architects were elbowed out of the competition, such as Giacomo della Porta, the municipal architect of Rome and architect of the Fabric of St. Peter. Interestingly Fontana himself addressed this issue in the first plate of his book in a quite ironic way [fig. 2]. At the base of the obelisk some of the competing

Fig. 2 Competing projects for the transport of the Vatican obelisk.

5 Belli, Gianluca, "Un monumento per Cosimo I de' Medici. La colonna della Giustizia a Firenze", in: *Annali di architettura*, vol. 16, 2005, 57-78.

architects are shown busy with models of their projects, while two *putti* descend from heaven with Fontana's own lifting machine. In the caption the depicted projects are explained only in terms of their functional principles; unfortunately for the historian the names of their respective inventors are not given. Marcello Fagiolo considered the model marked in the drawing with the letter "D" to represent Ammannati's project which proposed to transport the obelisk in an upright position.[6] Only model "B" can be identified with certainty, as it corresponds perfectly to a project already published by Camillo Agrippa in 1583.[7] Several of the other models also are highly reminiscent of ideas that had been circulating for some time. For example, model "E" with its inclined transportation stage has strong parallels to a project by Jacques Besson published around 1571/1572.[8] Furthermore, model "H" is based on much older ideas, as recently pointed out by Veronica Biermann.[9] The remarkable but rather impractical threaded bars of this project trace back to some well-known designs by Francesco di Giorgio Martini, who already in the fifteenth century devised various machines for moving and raising monolithic columns and obelisks.

In view of the various pre-existing projects which were copied or reworked in the context of the competition it might seem unsurprising that Domenico Fontana was criticised for having taken the ideas of others and claiming them to be his own inventions. For instance, Vincenzo Scamozzi refused to credit Fontana alone with the erection of the obelisk, referring not only to his competitors Ammannati and della Porta but also to some preceding architects such as Jacopo da Vignola and Antonio da Sangallo. These latter two had developed plans for the *guglia* project already in the first half of the sixteenth century.[10]

And indeed Lothar Sickel has recently produced evidence that such accusations were well-founded.[11] In a letter dated to February 1581 Camillo Paleotti, a legate to the Vatican, reported on a young Sicilian architect who by that time had applied to Pope Gregory XIII and the papal court with a project for the moving of the obelisk. The measures for lifting and transporting the obelisk that this architect – his name is unfortunately not mentioned – proposed and exemplified with a scale model are strikingly similar to the methods utilised by Fontana some years later. It is safe to assume that Fontana was aware of this project when he was commissioned with the transport of the obelisk in 1585.

6 Fagiolo, "Ammannati" (fn. 4), 195.

7 Agrippa, Camillo, *Trattato di Camillo Agrippa milanese di trasportar la guglia in su la piazza di San Pietro*, Rome 1583.

8 Besson, Jacques, *Theatrum instrumentorum et machinarum Iacobi Bessoni Delphinatis*, Lyon 1578; the book was first published around 1571/1572 in a small edition and then posthumously from 1578 onwards in various enlarged editions.

9 Biermann, Veronica, "Ortswechsel: Überlegungen zur Bedeutung der Bewegung schwerer Lasten für die Wirkung und Rezeption monumentaler Architektur am Beispiel des Vatikanischen Obelisken", in: Altekamp, Stefan/Marcks-Jacobs, Carmen/Seiler, Peter (ed.), *Perspektiven der Spolienforschung*, vol. 1, Spoliierung und Transposition, Berlin/Boston 2013, 133.

10 Scamozzi, Vincenzo, *L'idea della architettura universale*, Venice 1615, 337.

11 Sickel, Lothar, "Un progetto per il trasporto dell'obelisco vaticono descritto da Camillo Paleotti nel febbraio 1581", in: *Strenna dei Romanisti*, vol. 68, 2007, 653-662.

So, did Fontana simply steal the invention of the Sicilian architect? Do we have to perceive the nice motif of the two *putti* bringing along his lifting device just as a cynical representation of how the project came 'flown' to him?

2. A PROFESSIONAL REPUTATION UNDER CONSTRUCTION

In view of Fontana's occupational situation it is quite comprehensible that he always tried to promote his career as an architect. He had started out as mason and stuccoist and rose to the position of an architect by building the villa for Cardinal Montalto.[12] He must have felt lucky when the cardinal, his patron, was elected pope and could subsequently provide him with prestigious building projects. But in the Vatican it was all too well known that an architect who was favoured by a pope could easily fall from grace with the successor. Fontana was undoubtedly aware that his career depended almost exclusively on his patron Sixtus V and that he had to commend himself for a new position in good time.

However, Fontana was a master of self-promotion. The most important factor in his self-advertising was certainly his book which was published in 1590, the year Sixtus V died. But already several years earlier, Fontana seems to have used any opportunity and any medium to publicise his achievements, especially the moving of the obelisk. This started on the construction site of the *guglia* itself: Fontana could certainly reckon on a great public interest in the operation, but he still aimed to increase this attention. He ordered the site to be fenced off and had it protected by the Swiss Guard; on the day of the first lifting of the obelisk it was forbidden under penalty of death to enter the construction site or only to talk out loud.[13] In his report Fontana explained these measures with his concern that the course of action might be disturbed, but at the same time he obtained an almost theatre-like arrangement. Directed by acoustic signals from a trumpet and a bell, more than 900 workmen and 75 horses operated like a choreographed performance. Thousands of people, among them many clerical and secular dignitaries, witnessed the great spectacle. Fontana obviously had planned not only the operation but also its promotion in detail. On the evening of the first day, when the obelisk had been lifted successfully from its base, even the cannoneers of *Castel Sant'Angelo* knew what they had to do and concluded the day with gun salutes.

Moreover, well before the re-erection of the obelisk in St. Peter's Square was completed in September 1586, the event was commemorated in a large-sized engraving dated to August of the same year. It depicts the

12 Bedon, Anna, "Venture e sventure finanziarie del Cavalier Domenico Fontana", in: Fagiolo, Marcello (ed.), *Studi sui Fontana: Una dinastia di architetti ticinesi a Roma tra Manierismo e Barocco*, Rome 2008, 39.

13 Fontana, *Della trasportatione* (fn. 1), fol. 13 r.

construction site with great attention to detail [fig. 3] and also features Fontana in an inset with a portrait and a pompous heading that explicitly names him as architect and director of the project. The print was designed, engraved and edited by Giovanni Guerra (1544-1618) and Natale Bonifacio (ca. 1537-1592);[14] apparently it was intended to be sold as a souvenir for the visitors who came in large numbers to witness the spectacle. In 1586 and 1587 three more engravings dealing with the obelisk were produced by the same artists. All four prints were protected by a papal copyright, a fact that illustrates the close relationship of Domenico Fontana and his circle with Pope Sixtus V.

Fig. 3 The lifting of the Vatican obelisk observed by spectators on the rooftops of Old St. Peter's Basilica.

14 Witcombe, Christopher, *Copyright in the Renaissance: Prints and the privilegio in sixteenth-century Venice and Rome*, Leiden 2004, 276-282.

But this visual promotion continued: Guerra depicted the moving of the *guglia* in a fresco in a wing of the Vatican Library, newly erected by Domenico Fontana in 1588-89. In the same years the motif of the obelisk was likewise displayed on several frescoes in the Lateran Palace, another building project by Fontana.[15] Of course, these depictions were meant to praise primarily the initiator, i.e. the pope. In particular the fresco in the Vatican Library forms a part of the cycle depicting the good deeds of Sixtus V. But at same time the paintings increased Fontana's fame. This is evident by the fact that even in Fontana's private house a frieze of frescoes celebrated the raising of the obelisk together with his buildings for Sixtus V.[16]

Furthermore, the obelisk project was publicised in another very prestigious medium of that time. Various silver and gold medals minted in the name of Sixtus V praised the moving of the *guglia* or the re-erection of three other ancient obelisks conducted by Fontana from 1587 to 1589. Additionally, numerous bronze medals that immortalised not only the obelisks but also Fontana's portrait were produced – possibly on his own initiative.[17]

Along with the dynamic self-promotion induced by his contingent employment came Fontana's pride in his technical and architectural achievements. This led to another quite unusual form of self-representation: he literally signed some of his buildings. As noted by Christof Thoenes, the left gateway of the Royal Palace of Naples bears an inscription which declares that the designer was Domenico Fontana, Roman patrician and knight of the Golden Spur.[18] Other inscriptions also emphasize Fontana's social status, especially the knighthood that he was given by Sixtus V in recognition of the successful relocation of the Vatican obelisk. For instance, the signature on the base of the Lateran obelisk, the third and tallest obelisk re-erected by Fontana, names him as knight and architect. Only the inscription on the base of the Vatican obelisk refers to him simply as Domenico Fontana from the region Melide in the district of Como. Most probably it was already engraved before 26 September 1586, the day of the solemn consecration of the obelisk;[19] two days later, on 28 September, Fontana was given the knighthood of the Golden Spur.[20]

Along with the knighthood Domenico Fontana was granted a coat of arms. The heraldic figure that he chose referred to both his masterpiece and his name: an obelisk flanked by a fountain (*fontana*) on each side. Since

---

15 Mandel, Corinne, "Simbolismo ermetico negli obelischi e colonne della Roma Sistina", in: Fagiolo, Marcello (ed.), *Sisto V. Roma e il Lazio*, Rome 1992, 659-692 for the frescoes in the Vatican Library and in the Lateran Palace.

16 Massimo, Vittorio, *Notizie istoriche della Villa Massimo alle Terme Diocleziane con un'appendice di documenti*, Rome 1836, 90 and plate IV.

17 Sacchi, Ferdinando, "Domenico Fontana e le medaglie degli obelischi di Sisto V", in: *Emporium*, vol. 95, 1942, 121-126; Severin, Ingrid, *Baumeister und Architekten. Studien zur Darstellung eines Berufsstandes in Porträt und Bildnis*, Berlin 1992, 31-32, 181-182.

18 Thoenes, Christof, "Perché studiare Domenico Fontana", in: Curcio, Giovanna/Navone, Nicola/Villari, Sergio (ed.), *Studi su Domenico Fontana*, Mendrisio 2011, 9-19.

19 Fontana, *Della trasportatione* (fn. 1), fol. 33 verso wrongly dates the consecration to 27 September.

20 D'Onofrio, *Obelischi di Roma* (fn. 3), 90-91 and Perrault, Claude, *Les dix livres d'architecture de Vitruve, Paris* ^{2}1684, 339-342..

that time Fontana was commonly known as the knight of the obelisk, the *cavaliere della guglia*.

These various elements of self-promotion and self-representation are combined in the frontispiece of Fontana's book [fig. 4]. An opulently decorated *aedicula* frames a half-length portrait of Domenico Fontana, elegantly dressed with a mantle and a ruff. In a playful manner the *cavaliere* holds a small obelisk in his hands; additionally his right hand toys with a chain and a medal – certainly the golden chain that Sixtus V gave him as a present on the occasion of his ennoblement. On a desk in front of him he has placed the insignia of his profession, the square and the compass. The inscription running along the frame of the portrait attests that this is Domenico Fontana from Melide in the diocese of Como, architect of His Holiness, at the age of 46. His status as a knight is indicated by two sets of his coat of arms and helmets placed on the pedestals of the two columns framing the *aedicula*.

The frontispiece which again was engraved by Natale Bonifacio exemplifies another characteristic of Fontana's *modus operandi*: with a keen sense of artistic quality he recruited talented artists and invested in first class techniques in order to maximise the impact on his audience.

Fig. 4 Frontispiece depicting Domenico Fontana, 1590.

This begins with the famous model that he chose for his cover picture. As stated by Christof Thoenes, Fontana took up the tradition of the frontispiece from Jacopo da Vignola's *Regola delli cinque ordini d'architettura*.[21] The high graphical quality of Fontana's frontispiece also seems to compete with this model. On the whole, Fontana spared no expenses for his volume. While high-grade printing technology using copper engravings was used for all the plates – just like in Vignola's book –, this was in stark contrast to the majority of other sixteenth century treatises on architecture, which instead featured simpler and less expensive woodcuts, such as the works by Sebastiano Serlio or Andrea Palladio.

3. BREAKING GROUND FOR A DISCOURSE ON ENGINEERING

In his book, Domenico Fontana is presented primarily as an architect; roughly two thirds of the volume deal with the buildings that he realised for Sixtus V, while one third is dedicated to the moving of the obelisk. Nevertheless, the book became famous for the account on the transport of the *guglia*. This might be due to the gripping way in which the report was written – most probably Fontana did not write it himself but had a talented ghostwriter for this task.[22] Moreover, documentation describing the moving of a heavy load was a unique feature in contemporary technical literature, which brimmed with fantastic ideas and unrealistic projects.[23] Fontana obviously realised that the transport of the *guglia* had an enormous potential to attract public attention and he has to be credited with having exploited this potential thoroughly. In doing so he established that a technical operation was a memorable achievement that was worth being depicted and described in a book. It is not an exaggeration to say that Domenico Fontana created a new genre of technical literature: the work report that was based on facts. As we will see, the report on the Vatican obelisk formed an important point of reference in technical treatises and in engineering far into the nineteenth century.

4. THE DISCOURSE AND ITS DEVELOPMENT

In 1613/1614, when Carlo Maderno (ca. 1556-1629) erected the Marian column in front of *Santa Maria Maggiore* in Rome, he adopted basically the same technique that had been used for the Vatican obelisk.[24] This is not surprising since Maderno was the nephew of Domenico Fontana and had assisted

---

21 Thoenes, "Perché studiare Domenico Fontana" (fn. 18), 12-14.

22 Teichner, Richard, "Zur Biographie Domenico Fontanas", in: Conrad, Dietrich (ed.), *Domenico Fontana: Die Art, wie der Vatikanische Obelisk transportiert wurde*, Berlin/Düsseldorf 1987, 61.

23 Cf. for example Besson, *Theatrum instrumentorum* (fn. 8) or Ramelli, Agostino, *Le diverse et artificiose machine*, Paris 1588.

24 Marconi, Nicoletta, "Machine and symbol: between tradition in the execution and technical progress. The erection of the Marian column in Piazza Santa Maria Maggiore in Rome (1613-1614)", in: Dunkeld, Malcolm (ed.), *Proceedings of the Second International Congress on Construction History*, vol. 2, Cambridge 2006, 2077-2093.

him with the re-erection of all four obelisks a quarter of a century earlier.[25] He was familiar with the proven procedure and – being the architect of the Fabric of St. Peter – he even had some of the original tools and tackles available that had been stored in the magazines of the *Fabbrica*. In raising the Marian column – an ancient marble monolith taken from the Basilica of Maxentius – Maderno directly continued the tradition of Domenico Fontana.

But even as far away as Warsaw, and more than 50 years after the moving of the Vatican obelisk the book of Domenico Fontana produced an echo. Here, in front of the Royal Castle, a column in honour of King Sigismund III Vasa (1566-1632) was erected in 1643/1644; according to an engraving by Willem Hondius (ca. 1597-ca. 1652) this monolithic column was raised by means of a scaffold that had remarkable parallels to the one built by Domenico Fontana.[26] But as the Sigismund Column was much smaller than the *guglia*, the adoption of Fontana's method seems unnecessary. It rather indicates an emblematic imitation of the Roman model, although with a different symbolic intent. The four obelisks re-erected by Fontana as well as the Marian column raised by Maderno held a foremost religious meaning. Bronze crosses placed on top of each obelisk converted the ancient pagan objects literally into supports of the Holy Cross; correspondingly Maderno transformed the column into a trophy of the counter-reformation by placing a statue of the Virgin Mary at its top. The column in Warsaw, on the contrary, carried a statue of King Sigismund. The raising of this column quoted the Roman model as the epitome of a technical effort which was brought here into service for the royal dynasty of Poland. In a similar way, Hondius quoted Fontana's book, especially the mode of presenting technology, when he depicted the construction of the monument in his engraving.

Later in the seventeenth century engineering works gained an increasing significance as memorable events that were worth publicising. This is exemplified in the construction of the famous eastern façade of the Louvre Palace from 1667 onwards. In 1674 the central pediment was covered with two enormous stone slabs whose size exceeded every structural or functional need.[27] With a length of 16.90 m they were only 49 cm thick and therefore liable to break. The technical accomplishment of handling these giant slabs seems to have been the purpose *per se*. The architect Claude Perrault (1613-1688) and the master carpenter Poncelet Cliquin developed special wooden cages for transporting the slabs as well as an elaborate lifting device for putting them on top of the pediment. Ten years after the event Perrault presented this machinery in an accurate work report which he included in his second edition of Vitruvius.[28] The fact that in this account neither Domenico Fontana nor the Vatican obelisk are mentioned is not surprising since Perrault was one of the protagonists in a contemporary

25 Hibbard, Howard, *Carlo Maderno and Roman architecture 1580-1630*, London 1971, 36.

26 Chroscicki, Juliusz, "Sisto V e la Polonia: I rapporti politici ed artistici", in: Fagiolo, Marcello (ed.), *Sisto V. Roma e il Lazio*, Rome 1992, 851-863, esp. fig. 11.

27 Petzet, Michael, *Claude Perrault und die Architektur des Sonnenkönigs: Der Louvre König Ludwigs XIV. und das Werk Claude Perraults*, Munich/Berlin 2000, 316-319; see also Berger, Robert, *The palace of the sun: the Louvre of Louis XIV*, University Park 1993, 58-59.

28 Perrault, *Les dix livres d'architecture de Vitruve* (fn. 20), 339-342.

rivalry which was going on between Paris and Rome for the leading position in architecture. It is safe to assume that Perrault had a thorough knowledge of the moving of the *guglia* since he himself had been engaged in an obelisk project for Louis XIV in 1666.[29] Moreover, it is documented that Fontana's book circulated at the French court, since the most influential minister and advocate of the Louvre project Jean-Baptiste Colbert owned a copy.[30] In fact, some of the characteristics that Perrault described in his report clearly relate to Domenico Fontana's practice, for example the strict silence that had to be kept during the lifting of the slabs. Similarly the mode of presenting the technical apparatus recalls that of Fontana; the devices were displayed in illustrations not only in Perrault's account but also in an impressive engraving by Sébastien Le Clerc which showed the construction site in action.[31] However, these drawings were now focused mainly on technical aspects. In the third figure of his account Perrault even provided an alternative solution for the lifting machine which, in addition, was more sophisticated than the one he had actually used. In contrast to Fontana's self-glorifications, Perrault's and Le Clerc's representations paid only little attention to the personal achievement of the architect; they certainly implied that all these technical accomplishments served the greater glory of the Sun King.

A particular interest in technical details can also be observed in a new edition of Domenico Fontana's report on the obelisk that was published in 1694 by Carlo Fontana (1638-1714).[32] As noticed by Antonio Becchi there are some technical inconsistencies in the original drawings of the scaffolding.[33] In two plates of Domenico Fontana's book, fol. 12 r and 18 r, the top of the scaffold is depicted with rectangular trusses, whereas in fol. 20 r it is shown with a triangular *capriata,* i.e. a simple king post truss. Carlo Fontana obviously realised these discrepancies. As he had to reproduce all the engravings – the original copperplates had been purchased some decades earlier by the Dutch publisher Joan Blaeu – he tried to solve the logical conflict by modifying some of the drawings. For example, in his plate fol. 143, which corresponds to fol. 18 in the original report, he inserted a king post into the rectangular truss in order to bring it more in line with the other illustrations. But since there were some more contradictory details, for example in the lateral bracing, Carlo Fontana did not quite succeed in reconstructing a coherent idea of the scaffold in the different stages of the operation. All in all, he copied eight of the twelve original engravings with some slight modifications and added six new drawings, among them the famous panoramic view of the construction site on St. Peter's Square [fig. 1]. However, when Carlo Fontana dedicated the third book of his volume on the history of St. Peter's entirely to the episode of the *guglia* he was not only interested in matters of engineering. As Giovanna

29 Petzet, *Claude Perrault* (fn. 27), 335-353.

30 Mühlner, Manfred, "Fontanas Buch als opus curiosum et rarum", in: Conrad, Dietrich (ed.), *Domenico Fontana: Die Art, wie der Vatikanische Obelisk transportiert wurde*, Berlin/Düsseldorf 1987, 97.

31 Petzet, *Claude Perrault* (fn. 27), 318-319 and fig. 216.

32 Fontana, Carlo, *Templum Vaticanum et ipsius origo*, Rome 1694, esp. 107-173.

33 Becchi, Antonio, "Cantieri d'inchiostro. Meccanica teorica e meccanica chirurgica nella seconda metà del Cinquecento", in: Curcio, Giovanna/Navone, Nicola/Villari, Sergio (ed.), *Studi su Domenico Fontana*, Mendrisio 2011, 94.

Curcio pointed out, he likewise wanted to present himself as descendant of Domenico Fontana although they were only remote relatives.[34] By retelling the glorious deeds of his alleged ancestor he sought to enhance his own prestige.

Ten years later Carlo Fontana and his son Francesco (1668-1708) were confronted with a task comparable to the raising of an obelisk, when in 1703 the Column of Antoninus Pius was rediscovered in the ancient *Campus Martius* in Rome. The giant monolithic granite column, which had remained standing on its pedestal throughout the centuries, was at that time buried for half of its height and closely surrounded by buildings.[35] After the column had been excavated completely in 1704 Pope Clement XI decided to have it removed from its narrow pit and re-erected in *Piazza di Monte Citorio*. In a committee of experts Carlo and Francesco Fontana proposed to lift the column by means of a scaffold that – of course – corresponded closely to the one designed by Domenico Fontana.[36] When Francesco was commissioned with the project he continued precisely according to Domenico's book. The construction site was fenced in and protected by the Swiss Guard, with the workmen strictly forbidden to speak out loud since they were to be directed by signals from a trumpet and a bell. Even some of the original tackles could be used since they were still kept in the Fabric of St. Peter.

But the operation turned out to be complicated, since the construction site was very narrow and the eight capstans for the lifting of the column could not be equally spaced. Furthermore, on 26 September 1704, that is already before the column was moved, two cracks had been noticed in the shaft and thus it was reasoned that the column was broken into three pieces.[37] As a precaution, a test for the lifting was arranged on 15 October which proved satisfactory. The column could be raised some centimetres even though one of the horizontal beams on top of the scaffold failed under the load. However, the main event which was scheduled for 18 October took a less successful course. In the presence of many spectators, among them the queen of Poland and numerous dignitaries and noblemen, various accidents happened. First another one of the horizontal beams gave way and had to be fixed, with a second interruption caused by the breakdown of a capstan. When finally one of the trusses of the scaffold broke apart the operation had to be called off.

Repairs to the scaffolding took quite a long time; the competence of Francesco Fontana was under discussion and Clement XI therefore felt the need to confirm him as the engineer in charge. Almost a year later, on 24 September 1705 Fontana made another attempt and this time he wanted to play it safe: the scaffolding had been reinforced, there were now thirteen instead of eight capstans, and the number of workmen was raised from 340 to 520.[38] In the end the column was lifted from its pit without difficulty. A

34 Curcio, Giovanna, "Del Trasporto dell'Obelisco Vaticano, e sua Erezione", in: Curcio, Giovanna (ed.), *Il Tempio Vaticano 1694. Carlo Fontana*, Milan 2003, CLXX.

35 D'Onofrio, *Obelischi di Roma* (fn. 3), 238-249 for the column and its lifting.

36 Marconi, Nicoletta, "Uso delle macchine da costruzione e 'crisi' della capacità operativa nel XVIII secolo: il fallimento del trasporto della colonna di Antonino Pio a Roma", in: *Quaderni dell'Istituto di Storia dell'Architettura*, vol. 33, 1999, 43-54.

37 Marconi, "Uso delle macchine" (fn. 36), 48.

38 Ibid., 47.

contemporary engraving by Arnold van Westerhout which was based on a drawing by Francesco Fontana depicts this moment of success.

Despite these efforts, the Column of Antoninus Pius was never re-erected. After it had been brought to *Piazza di Monte Citorio* it was left there because the damages it had suffered over the centuries turned out to be rather serious and made an immediate re-erection impossible. More than 50 years later the column was still lying there when Giovanni Battista Piranesi addressed its history in his volume on the ancient *Campus Martius*.[39] Together with a reprint of van Westerhout's engraving Piranesi published two plates that illustrate the state of preservation of the column and its marble pedestal.[40] The shaft shows deep cuts and grooves that would have required a substantial restoration, but apparently the necessary red Aswan granite was scarce. Finally, in 1788 the column was cut into pieces in order to use its stone for the restoration of an ancient obelisk that had been found nearby and that was erected in *Piazza di Monte Citorio* instead of the column.[41]

Modern research has rated the column project as a failure caused by mistakes on the part of Francesco Fontana, and has raised questions whether his scaffold was the adequate device for this task.[42] However, a look at several contemporary sources would suggest a less critical evaluation. For example, a statement by Francesco Bianchini, one of the leading scholars in Rome around 1700 and member of the committee concerned with the column, came to the conclusion that Fontana's scaffold was very safe and rather blamed the carpenters for having assembled it incorrectly.[43] This would indicate that the method applied by Domenico Fontana in 1586 for the Vatican obelisk was still state of the art in 1705. In fact, it would take more than another century to develop a new technical paradigm for such projects.

5. THE DISCOURSE IN TECHNICAL TREATISES

The knowledge of Domenico Fontana's method for transporting the Vatican obelisk was spread first by his book which was published in two editions: a first one issued in 1590 by Domenico Basa in Rome and in a second rarer edition published in 1604 by Costantino Vitale in Naples.[44] Later on, the report on the moving of the *guglia* was summarised in various books and thereby became more widespread, for example in the third volume of Athanasius Kircher's *Oedipus Aegyptiacus* in 1654.[45] In 1663 Joan Blaeu published a Latin

39 Piranesi, Giovanni Battista, *Il Campo Marzio dell'Antica Roma*, Rome 1762, plates XXXI-XXXIII.

40 See Wilton-Ely, John, *Giovanni Battista Piranesi: The complete etchings*, vol. 2, San Francisco 1994, 647-649, esp. 648, no. 593.

41 D'Onofrio, *Obelischi di Roma* (fn. 3), 288-289.

42 Marconi, Nicoletta, "L'eredità tecnica di Domenico Fontana e la Fabbrica di San Pietro: tecnologie e procedure per la movimentazione dei grandi monoliti tra '500 e '800", in: Fagiolo, Marcello (ed.), *Studi sui Fontana: Una dinastia di architetti ticinesi a Roma tra Manierismo e Barocco*, Rome 2008, 45-56, esp. 52; see also Marconi, "Uso delle macchine" (fn. 36); D'Onofrio, *Obelischi di Roma* (fn. 3), 243-246.

43 Marconi, "Uso delle macchine" (fn. 36), 48-49.

44 Carugo, Adriano, "Notizia bibliografica", in: Carugo, Adriano (ed.), *Domenico Fontana: Della trasportatione dell'obelisco Vaticano 1590*, Milan 1978, LXI-LXIV.

45 For the treatment of Domenico Fontana in artistic biographies and early art history see Caraffa, Costanza, "Domenico Fontana e gli obelischi. Fortuna critica del 'Cavaliere della Guglia'", in: Curcio, Giovanna/

translation of the report that he supplemented opulently with prints produced from the original copper plates.[46] Blaeu included the account in the second volume of his atlas of Italy in order to illustrate the glorious past of ancient Rome; obviously the episode of the Vatican obelisk had become an integral part of the history of the Eternal City.

Apart from such publications that were motivated above all by an interest in history, the report on the obelisk was increasingly featured in treatises and books dealing with technical topics and engineering. Jacob Leupold (1674-1727) for instance, discussed two different ways to erect an obelisk in his treatise on lifting devices which was published in 1725. One of them – from today's perspective quite unrealistic – proposes to raise the obelisk with a multitude of screws and to gradually push it into an upright position, whereas the other one is Domenico Fontana's method [fig. 5, 6].[47] Leupold did not know Fontana's report in its original version, and he complained that he could not obtain it anywhere, although he had searched for it intensively for many years. However, it seems that he knew the illustration of the scene from Athanasius Kircher's account and reproduced this version; at the same time he criticised it because of an

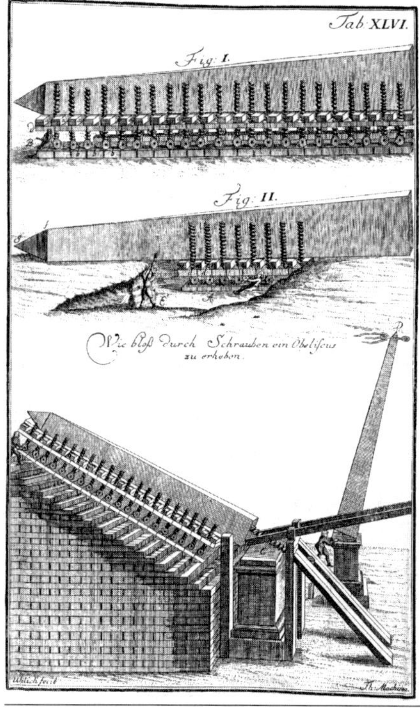

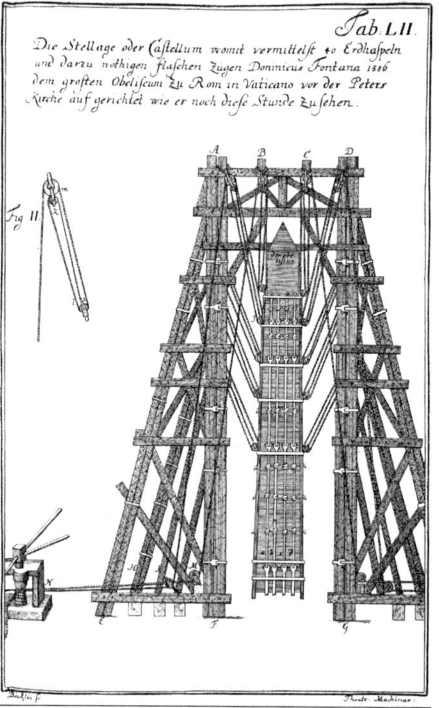

Fig. 5 Method of raising an obelisk by means of screws.
Fig. 6 Domenico Fontana's method of lifting the Vatican obelisk.

Navone, Nicola/Villari, Sergio (ed.), *Studi su Domenico Fontana*, Mendrisio 2011, 21-47.

46 Blaeu, Joan, *Theatrum Civitatum et Admirandorum Italiae*, vol. 2, Amsterdam 1663, 271-315.

47 Leupold, Jacob, *Theatrum Machinarium, Oder: Schau-Platz der Heb-Zeuge*, Leipzig 1725, 128-129 and plate XLVI and respectively 137-145 and plate LII.

unclear detail in the depiction of the protective casing of the obelisk. Leupold also commented on the scaffold which Francesco Fontana used to lift the Column of Antoninus Pius, in this case citing van Westerhout's engraving as his source.[48] In comparison with the lifting machines of both Fontanas, which functioned simply by means of tackles and capstans, Leupold seems to have been more fascinated by the technical implications of the other lifting device which worked with screws. However, in his final appraisal of the Fontanas' approach Leupold argued against a potential criticism pointing out that their method was very practical and that it worked quite efficiently.

In the eighteenth century Domenico Fontana's lifting device had thus become an epitome for a large-scale project among specialists, like technicians and engineers. This is exemplarily demonstrated by Johann Jacob Schübler's treatise on carpentry published in 1731, which featured Fontana's scaffold with the obelisk on the frontispiece.[49] Another example is the treatise *Castelli e Ponti* published in 1743 under the name of Nicola Zabaglia which presented the many elaborate scaffoldings that this craftsman had developed for the Fabric of St. Peter.[50] Additionally, the transport of the *guglia* is here described once again and illustrated with prints produced from Carlo Fontana's copper plates – not least in order to praise the glorious achievements of the Fabric of St. Peter and to present it as Europe's leading workshop for developments in engineering.[51]

In one particular case the Vatican obelisk and Domenico Fontana's lifting device even became a topic in the arts when the famous frieze of the "Art of War", which Francesco di Giorgio Martini had designed originally in the fifteenth century for the Ducal Palace in Urbino, was newly assembled and enlarged in 1756. Giovan Francesco Buonamici (1692-1759) added 17 new reliefs to this series which represented the "Machines of Peace", i.e. important inventions of civil engineering such as the tackle or the water wheel. Two of these reliefs show the Vatican obelisk and its scaffold and thereby illustrate the importance that was assigned to Domenico Fontana's achievement in the eighteenth century [fig. 7].

In the nineteenth century the moving of the Vatican obelisk still formed a regular part of the treatises on engineering. In addition to a second edition of *Castelli e Ponti* published in 1824, a number of contemporary French treatises should be mentioned: Jean-Baptiste Rondelet's famous *Traité théorique et pratique de l'art de bâtir* published between 1802 and 1817 as well as a treatise on the lifting of heavy loads, that was edited by Joseph-Antoine Borgnis in 1818, and a treatise on carpentry by Amand-Rose Emy from 1841.[52] However, in these publications the transport of the Vatican

48 Leupold, *Theatrum Machinarium* (fn. 47), 146-148 and plate LIII.

49 Schübler, Johann Jacob, *Nützliche Anweisung Zur Unentbehrlichen Zimmermanns-Kunst*, Nuremberg 1731.

50 Zabaglia, Nicola, *Castelli e ponti*, Rome 1743; esp. 12-21 and plates 37-52 for the moving of the Vatican obelisk.

51 Marconi, Nicoletta, "La 'prestigiosa collazione delle macchine del Zabaglia' e la 'scuola' di meccanica pratica della Fabbrica di San Pietro", in: Marino, Angela (ed.), *Sapere e saper fare nella Fabbrica di San Pietro. Castelli e ponti di maestro Niccola Zabaglia 1743*, Rome 2008, 54-82.

52 Rondelet, Jean-Baptiste, *Traité théorique et pratique de l'art de bâtir*, Paris 1802-1817, plate 170; Borgnis, Joseph-Antoine, *Traité complet de mécanique appliquée aux arts. Mouvemens des fardeaux*, Paris 1818,

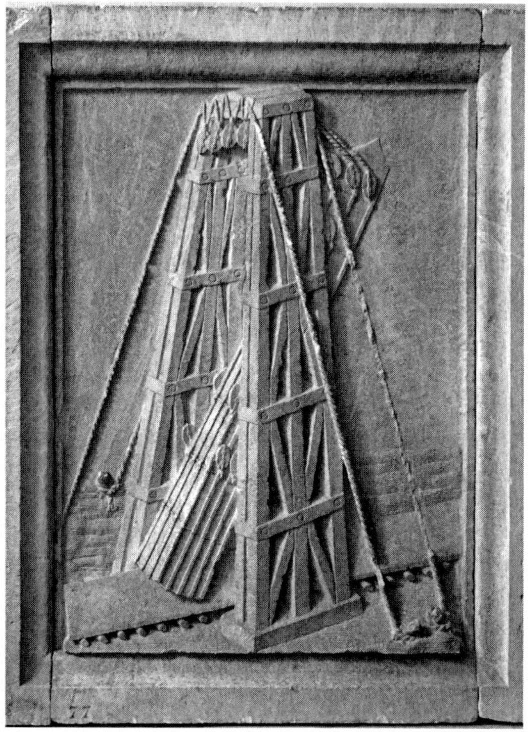

Fig. 7 Relief depicting the transport of the Vatican obelisk (Giovan Francesco Buonamici, 1756).

obelisk had lost its significance as the most noteworthy achievement in engineering. Other mile stones in the history of engineering were now treated on equal level with the *guglia*: the pediment stones of the Louvre Palace, the Column of Antoninus Pius, the relatively small Roman obelisk of Arles re-erected in 1676, as well as the so-called Thunder Stone of Saint Petersburg, to which we will return shortly. These French treatises did not engage in depth with Domenico Fontana's lifting device nor did they discuss it in regard of its technical features, but instead re-stated the well known facts and placed the project in an assembly of major engineering events in a quasi-encyclopaedic fashion. However, these treatises and their various later editions kept the memory of the moving of the Vatican obelisk alive well into the later nineteenth century.

6. THE RISE OF A NEW TECHNICAL PARADIGM

For engineers that had to transport heavy loads and above all large monoliths, the Vatican obelisk remained the point of reference until the early nineteenth century. This also applies to projects which from a technical perspective were comparable only to a limited extent, like the transportation of the

189-206, 312-324 and plate 13; Emy, Amand-Rose, *Traité de l'art de la charpenterie*, Paris 1841, 625-641, esp. 633-634 and plate 156.

so-called Thunder Stone of Saint Petersburg in 1769/1770. For this undertaking a huge granite boulder destined to form the base of the bronze equestrian statue of Peter the Great was brought to Saint Petersburg partly overland and partly by ship from Finland. This boulder, which was many times heavier than the Vatican obelisk, did in contrast not have to be lifted but was moved only by pulling. For this purpose the engineer in charge, Marinos Charvouris (1729-1782), developed a system of ball bearings over which the rock was moved horizontally. The loading of the boulder on board of a cargo ship presented another new technical challenge. Despite these obvious differences to the transport of the Vatican obelisk, Charvouris published a detailed report in 1777 which corresponded to the literary genre established by Domenico Fontana nearly two centuries earlier. He described all the technical aspects of his undertaking and illustrated the proceedings with engravings.[53] Interestingly, Charvouris referred to the project of the Vatican obelisk but obviously did not know about it from Fontana's book, since he wrongly believed that the *guglia* had been brought to St. Peter's Square from *Piazza Navona*. Even though in this case Fontana's account seems to have been known only indirectly from another publication, it served as a role model insofar as a technical achievement, the moving of a heavy load, was once again memorised as a noteworthy event and formed the main topic of a treatise. Charvouris' report suggests that the enormous size of the rock was one of the deciding features already during the planning phase of the project. The argument was brought forward that the technical *tour de force* of transporting one massive stone instead of several smaller ones would enhance the magnificence of the monument for Peter the Great. This line of thinking corresponds with another aspect: Charvouris mentioned that some persons in high positions had prompted him to publish his work report.[54] Apparently they considered a full description of the making of the monument to contribute to its splendour. In this way the publication formed a part of the monument itself.

From a technical perspective two projects of the 1830s mark a turning point in the engineering discourse here discussed: the erection of the Alexander Column in Saint Petersburg in 1832 and the relocation of the Luxor obelisk to Paris in 1836. The set up used to mount the honorary column of the Russian emperor Alexander I still corresponded in detail to the one used by Domenico Fontana. The architect Auguste de Montferrand (1786-1858) and Antonio Adamini (1792-1846), who was responsible for the lifting devices, constructed a scaffold according to the well-established model.[55] In this case everything had to happen on a larger scale, but this was no obstacle as enormous means were at their disposal. The colossal column shaft, a granite monolith that was about as long as the Vatican obelisk but twice as heavy, was brought by ship from Finland to Saint Petersburg. Above

53 Charvouris, Marinos, *Monument élevé a la gloire de Pierre-le-Grand*, Paris 1777.

54 Ibid., 3.

55 Navone, Nicola, "Antonio Adamini e l'elevazione della Colonna Alessandrina a San Pietroburgo", in: Navone, Nicola/Tedeschi, Letizia (ed.), *Dal mito al progetto: La cultura architettonica dei maestri italiani e ticinesi nella Russia neoclassica*, vol. 2, Mendrisio 2003, 697-711.

all, the engineers had an abundance of construction timber available; since the column had to be raised onto a tall pedestal, they constructed a gigantic ramp of wooden beams on the square in front of the Winter Palace, with the scaffold tower rising high above it. In 1836 de Montferrand published a report about the project – with an explicit reference to Domenico Fontana and the Vatican obelisk – and illustrated it with tables that emphasised the enormous dimensions of the construction site.[56]

Almost contemporary with the erection of the Alexander Column, the project of the Paris obelisk struck a new path in monolith transportation. The engineer Apollinaire LeBas (1797-1873) was the first who broke with Fontana's established method and employed a much simpler technique. Already the laying down of the obelisk in Luxor in 1831 was accomplished without lifting but instead by tipping the monolith over in a controlled manoeuvre.[57] Unlike Domenico Fontana, whose main concern had been to prevent the breaking of the *guglia*, LeBas put his trust in the stability of the wooden casing and the strength of the stone. With this method the machinery could be kept quite simple. It consisted basically of a tiltable cable-stayed pylon that held the obelisk during the laying down process slowly following it in its tilting movement. With the same apparatus the obelisk was re-erected in Paris on 25 October 1836 in only a few hours. While LeBas deviated from Fontana as a role model with this simple and efficient machinery, the staging of the re-erection as a public event followed again the known pattern. In the presence of King Louis Philippe and in front of an enormous crowd of more than 200.000 spectators the obelisk was set up on *Place de la Concorde*. At first it was planned that the spectacle would be augmented by a special effect: a steam engine was supposed to power the lifting device so that the obelisk would "rise majestically into the space without the help of any animal force"[58]. But the steam engine turned out to be too weak and had to be replaced by the well-known capstans and 350 workmen that were once again directed by the signals from a trumpet. Nevertheless, the event was subsequently featured in various paintings and numerous articles.[59] In 1839 LeBas himself published an extensive report, where – almost in the style of an adventure novel – he told the complete story of the obelisk's transport, from the expedition in Egypt to the re-erection in Paris.[60] He additionally dedicated an entire chapter to his forerunner Domenico Fontana, whose moving of the Vatican obelisk he described in great detail and implicitly contrasted it with his own method. He willingly came to Fontana's defence against the contemporary criticism concerning the supposedly unneeded effort for his huge scaffold. LeBas reminded his reader of the fact that Fontana belonged to an era when mechanics was still in its infancy and

56 De Montferrand, Auguste, *Plans et détails du monument consacré à la mémoire de l'Empereur Alexandre*, Paris 1836.

57 LeBas, Apollinaire, *L'obélisque de Luxor. Histoire de sa translation a Paris*, Paris 1839, for the transport of the obelisk and its re-erection in Paris.

58 LeBas, *L'obélisque de Luxor* (fn. 57), 155: "C'eût été un spectacle bien imposant que de voir un fardeau de 500 milliers s'élever majestueusement dans l'espace, sans le secours d'aucune force animale".

59 See Solé, Robert, *Le grand voyage de l'obélisque*, Paris 2004, 280-284.

60 LeBas, *L'obélisque de Luxor* (fn. 57).

expressed the wish to re-establish Fontana in "toute sa gloire"[61]. Of course, those critical remarks against Fontana could also be understood as a dig at the conspicuous project of the Alexander Column, which was published in Paris in the same year when the Luxor obelisk was re-erected.

7. THE END OF THE DISCOURSE

In the context here examined, LeBas' lifting device can be considered a turning point mostly because of its substantial difference from Fontana's model and not so much because of its technological innovations. In contrast to what might be expected from a machine of the nineteenth century, its main structure did not consist of iron but of wood, not to mention the episode of the inadequate steam engine. The new features of the machine age were adopted for the transport of monoliths only more than 40 years later with the so-called Needles of Cleopatra, the obelisks of London and New York. These two monuments taken from Alexandria in 1877 and 1879 respectively were lifted by means of hydraulic jacks instead of the tackles and capstans previously used. While the re-erection of the London obelisk in September 1878 was still realised with a wooden scaffold, the New York obelisk was supported solely by two iron frames when it was set up in Central Park in January 1881.[62] In both cases the obelisks were mounted on pivots that were fixed to the sides of the shafts close to their centre of mass; thus they were much more easily turned into an upright position than the Paris obelisk whose hinge had been placed at its bottom.

Another significant difference to all earlier projects is that the transport of both of Cleopatra's Needles was initiated and financed by private sponsors, and that the execution was in the hands of engineers who functioned also as general contractors. As a consequence, the relocation of the London obelisk that had met with various difficulties and incidents ended in a financial fiasco that the responsible engineer and entrepreneur John Dixon had to bear. For this reason it does not come as a surprise that Dixon did not publicise his undertaking in a book – although during the transport of the obelisk he had sought public attention with some newspaper articles.[63] Instead, James Alexander, who had started the initiative to bring the obelisk from Egypt to London but then felt elbowed out by Dixon, published the project in a book in order to present the enterprise from his point of view.[64]

A quite different situation occurred with the transportation of the second of Cleopatra's Needles to New York; this project was successfully accomplished and subsequently published by the engineer Henry Gorringe

61 LeBas, *L'obélisque de Luxor* (fn. 57), 172.

62 Gorringe, Henry Honychurch, *Egyptian obelisks*, London 1885, 96-109 for the London obelisk and 1-58 for the New York obelisk.

63 Iversen, Erik, *Obelisks in exile*, vol. 2, Copenhagen 1972, 90-147, esp. 124-136.

64 Alexander, James Edward, *Cleopatra's Needle. The obelisk of Alexandria: its acquisition and removal to England*, London 1879.

(1841-1885).[65] Apart from the technical innovations already mentioned there are still many elements in his project which recall the *modus operandi* of Domenico Fontana, especially the handling of the construction site as a public event. For instance, the site had to be protected from the gathering spectators already during the laying down of the obelisk in Alexandria [fig. 8]. Instead of the Swiss Guard that had shielded the Vatican obelisk, Gorringe arranged for a cordon of sailors of the Imperial Russian Navy whose flagship was at that time moored in the harbour of Alexandria. At the same time he tried to give the spectacle the appearance of national importance by having the star-spangled banner hoisted on top of the obelisk. Later in New York there were different public events to celebrate the acquisition of the obelisk. The ceremony of the laying of the cornerstone on 9 October 1880 was inaugurated with a parade of nearly 9.000 freemasons and a speech by their grandmaster. On the occasion of the re-erection of the obelisk, which took place on 22 January 1881, about 10.000 spectators gathered in Central Park despite the bitter winter cold. In the presence of the U.S. Secretary of State and the Secretary of the Navy the crowd witnessed how the obelisk was turned into its vertical position. Thanks to Gorringe's modern techniques the operation took only five minutes; it was concluded with the playing of the "national airs"[66] while the guard of honour, a battalion of marines, presented arms. Gorringe described all these details in his elaborate work report which is illustrated not only with

Fig. 8 The laying down of the obelisk in Alexandria in 1879.

65 Gorringe, *Egyptian obelisks* (fn. 62).

66 Ibid., 47.

drawings but also with numerous photographs. Furthermore, all preceding obelisk projects were recapitulated: three chapters that were co-authored by Seaton Schroeder retold the stories of the obelisks of Paris, London, and the Vatican – forming an indirect comment on the rise of New York among the world's most prominent cities.

The New York obelisk and Gorringe's book mark the end of the discourse on the art of moving monoliths which Domenico Fontana had started some 300 years earlier. For the last time the transport and the erection of an ancient obelisk was staged as a major public event and subsequently presented in a book by the responsible engineer.

Certainly, also in the twentieth century there were large monoliths to be transported. Especially in Rome there apparently existed a continuous line of projects that still adhered to the technique established by the local genius Domenico Fontana: from the Column of the Immaculate Conception erected in 1857 and the columns set up in the 1890s in the atrium of Saint Paul's Outside the Walls to the reconstruction of some ancient columns of the *Basilica Ulpia* in the Forum of Trajan in 1932.[67] However, none of these interventions was published; apparently they were no longer regarded as outstanding technical achievements. Even one project whose main purpose was propagandistic was never presented in a book, namely the Mussolini Obelisk that was set up in 1932 in Mussolini's Forum in Rome.[68] The erection of this monolith of Carrara marble for which a special futurist scaffold of reinforced concrete had been built was broadcasted in the new medium of that time and therefore belongs to another discourse: the presentation of engineering achievements in film and newsreel.

LIST OF FIGURES

Fig. 1: Fontana, Carlo, *Templum Vaticanum et ipsius origo*, Rome 1694, fol. 169.

Fig. 2: Fontana, Domenico, *Della trasportatione dell'obelisco vaticano et delle fabbriche di Nostro Signore Papa Sisto V*, Rome 1590, fol. 8.

Fig. 3: Detail from: Guerra, Giovanni/Bonifacio, Natale, *Disegno, nel quale si rappresenta l'ordine tenuto in alzar la Guglia*, engraving, August 1586.

Fig. 4: Fontana, Domenico, *Della trasportatione dell'obelisco vaticano et delle fabbriche di Nostro Signore Papa Sisto V*, frontispiece, Rome 1590.

Fig. 5: Leupold, Jacob, *Theatrum Machinarium, Oder: Schau-Platz der Heb-Zeuge*, Leipzig 1725, plate XLVI.

Fig. 6: Leupold, Jacob, *Theatrum Machinarium, Oder: Schau-Platz der Heb-Zeuge*, Leipzig 1725, plate LII.

Fig. 7: From the series "Machines of Peace" by Giovan Francesco Buonamici, 1756, Ducal Palace, Urbino (photograph by the author).

Fig. 8: Gorringe, Henry, *Egyptian obelisks*, London 1885, plate VIII.

67 Marconi, "Machine and symbol" (fn. 24), 2088-2091.

68 D'Amelio, Maria Grazia, *L'obelisco marmoreo del Foro Italico a Roma*, Rome 2009.

BUILDING THE KIEL CANAL

CONSTRUCTING A WATERWAY, AN UNDERSTANDING OF TECHNOLOGY AND A HARMONIC GERMANY (1886-1895)[1]

Eike-Christian Heine

In November 1889 the German Reichstag debated the construction of the Kiel Canal [fig. 1]. The conservative representative Conrad Graf von Holstein (1825-1897) had been part of a delegation that had toured the building site. He was still impressed. Through a moor the canal could only be dug with difficulty, he reported, because "the mushy mass of the ground would flow back together" and destroy everything. The engineers overcame such material defiance only with great effort, von Holstein continued. To stabilise the tracks on which railroads hauled material in large quantities, four rows of 13 metre long piles were driven in the ground and "connected by wire, to produce a somewhat stabile fundament." He recognised the formidable deployment of machinery, the "excavators, the railroad tracks, the railroad waggons, the locomotives, the machine shops, the buildings [...], there is a gigantic value behind all of this." Talking about the housing of the workers, he was completely satisfied with the condition: "The arrangement of the barracks is [...] exemplary." He closed his remarks, saying that "there has developed a good relationship between workers and employers."[2]

Von Holstein's words are typical for the interpretation of the canal by the majority of Germany's conservative political elite and bourgeois public: The engineering problems were discussed in detail, people struggled to

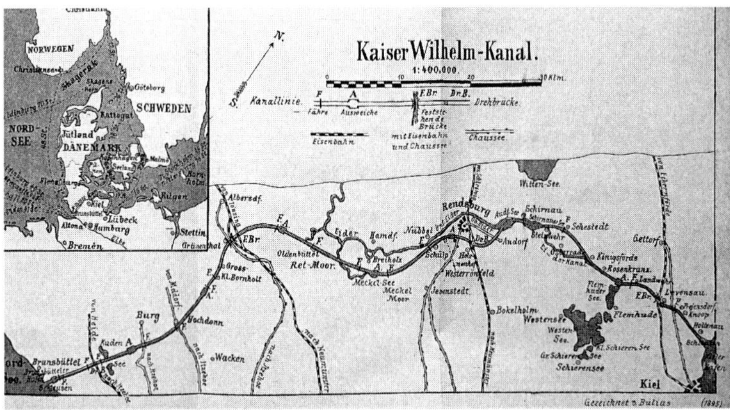

Fig. 1 Map of the Kiel Canal, 1895.

1 The article largely builds on my dissertation: Heine, Eike-Christian, *Vom großen Graben. Die Geschichte des Nord-Ostsee-Kanals*, Berlin 2015. Some of the aspects analysed there are also in: Ibid., "Connect and Divide. On the History of the Kiel Canal", in: *Journal of Transport History*, vol. 35, 2014, 200-219.

2 Graf von Holstein's address to the Reichstag on 20 November 1889, http://www.reichstagsprotokolle.de/150_Blatt3_k7_bsb00018661_00451.html (accessed 2-4-13).

understand the implication of industrial construction, and the social policy was praised repeatedly. This paper examines both material interventions into the landscape as well as wider imaginations that were under construction when the Kiel Canal was realised.

The Kiel Canal was opened with great festivities in 1895 and is still busy today. It cuts through Schleswig-Holstein, which is located on Jutland. At the bottom of this peninsular is the old trading port of Hamburg on the banks of the river Elbe. Schleswig-Holstein's history is deeply entwined with Denmark, which lies northwards and stretches until it ends at the Skagerrak. Any large ship had to pass this stormy sea if it wanted to navigate between the North Sea in the West and the Baltic in the East. The dukedom Schleswig-Holstein was for centuries ruled by the king in Copenhagen. Close to the royal palace at the gate to the Baltic every trader who passed between the seas was taxed. With the rise of nationalism and two wars between Germany and Denmark in 1848 and 1864, Schleswig-Holstein finally became a Prussian province in 1867. At the turn of the years 1885 and 1886, the German Reich decided to put its new wealth and the powers of the industrial age to use to build the Kiel Canal. This waterway was 100 kilometres long, 9 meters deep and 65 meters broad. Only a lock at Kiel on the Baltic and one close to the mouth of the Elbe kept the tides out of this monumental infrastructure project [fig. 2]. In 1895 passage was possible for the biggest vessels of that time. If one wants to be solemn, one can call it an artificial, manmade Skagerrak.

The comparison is justified in so far, as the economic and military debates about Denmark's control over the entrance to the Baltic now also determined the value of the canal. Government, the Reichstag and an enthusiastic German public agreed on two things. They were sure that the waterway would strengthen the navy that could transfer ships between the seas independently from the situation in the Danish straits. They were also sure that a canal would open the Baltic to world trade and made exchange

Fig. 2 The Kiel Canal is a deep cut through the landscape.

between German harbours easier by drastically reducing distances. Things turned out to be more complicated, however. Towns like Husum on the North Sea or Lübeck on the Baltic feared a watercourse because it would drastically strengthen Hamburg's port.[3] In many ways this seems to be the case until today, where Hamburg is the staple market for the Baltic. Also the canal's strategic relevance was controversial. When the cabinet debated the waterway's construction, it were only representatives of the military who raised objections. To them a canal seemed a strategic burden because it had to be secured by regiments of the army.[4] The canal also did not resolve the question of the Danish straits and the access of German enemies to the Baltic.[5] Both the dominance of Hamburg in trade and the question of military control of the canal and over the Belts turned out to be very real in the decades after the canal's completion.[6]

While these findings put a dent in the seemingly obvious reasoning of military and national economic benefits, the interpretations of the construction site were remarkably independent from any of these debates and impacts. In a first step this paper reconstructs some of the challenges of the building phase, which were the result of the landscape's materiality. The second chapter will explore how the construction site represented ideas of national and human progress as well as the contemporaries' struggle to understand technological change. The third step will investigate the modes in which the realisation of the Kiel Canal was perceived as the realisation of a better society through social policy. When the Kiel Canal was under construction, contemporaries were simultaneously building their understanding of technology and a vision – although a failed one – of a just class society.

1. SHAPING THE MATERIAL WORLD

To build a canal means to construct a reliable border that separates water inside the canal from its solid surroundings. In that respect the Kiel Canal was not different from other waterways. Easy as this task might sound, it had proofed difficult more than once, just think of the massive landslides that destroyed great parts of the French Panama Canal project.[7] The condi-

3 For Husum: *Zur Beleuchtung der Canal-Linie zwischen Eckernförde und Husum, von einem Ausschuß-Mitgliede*, Schleswig 1848; for Lübeck: "Gutachterliche Äußerung der Handelskammer zum Senats-Decrete v. 18. Juni 1864, die Anlage eines Kanals von der Nordsee nach der Ostsee betr." (Stadtarchiv Lübeck ASA Int. 28805).

4 E.g. Proceedings of the Prussian Staatsministerium, 20 December 1883 (Bundesarchiv R/901/7039).

5 Historical research about the Kiel Canal has for large parts overlooked this fact. Hans Branner has shown that Denmark remained a strategic element of uncertainty for the German military: *Småstat mellem stormagter, Belutningen om mineudlægning August 1914*, Copenhagen/Aarhus 1972, 63. For the contemporary debate see e.g. Anning-Petersen, P.C., "Nord-Østersøkanalen", in: *Militært Tidskrift* 1894, 275-288 or Bruchhausen, Karl von, "Die Neutralisation Dänemarks", in: *Preußische Jahrbücher*, vol. 78, 1897, 217-229.

6 Hamburg soon concentrated German Baltic trade in its harbour, see e.g. Gloe, Hans-Otto, "Der Kaiser-Wilhelm-Kanal", in: *Handel und Wandel am Kaiser-Wilhelm-Kanal*, Kiel 1929, 3-18. Even after the construction of the Canal, the Danish straits continued to be of strategic importance for the German military who now feared a British sneak attack on its harbours, see e.g. Epkenhans, Michael, "Germany and Denmark before 1914", in: Epkenhans, Michael/Groß, Gerhard P. (ed.), *The Danish Straits and German Naval Power 1905 1916*, Potsdam 2010, 3 12; Heine, *Vom großen Graben* (fn. 1), 220-232.

7 E.g. Parker, Matthew, *Panama Fever, The epic story of the greatest human achievement – The building*

tions were of course not as extreme as on the equator, but along the line of the Kiel Canal the surroundings proofed defiant enough when fine sands flowed into the construction pit and behaved rather like a liquid than a solid element. On other occasions groundwater penetrated the pit, hindered the shovelling of workers and dredgers, but also washed out ground-material and thus destructed banks. Two examples depict the difficulties, the soft ground of Schleswig-Holstein had in store for engineers, dredgers and workers.

In the Western part of the canal, south of the lake Kudensee, the canal passes through a depression. The swampy ground here was unable to support the dams that were needed to secure the water level in the canal one meter above its surroundings. Material thrown on the moor slowly disappeared in the soft soil. To build the dykes, sand was hauled in by rail and deposited until it completely replaced the old earth [fig. 3]. Now the banks had a stable fundament and the canal could be dug out. In retrospect one of the chief engineers labelled the 12 kilometres through the depression as the most difficult section. In order to stop the rails from sinking into the swamp, pairs of logs up to 13 metre long were rammed into the soil and connected with a net of metallic wire.[8] This installation served its purpose, but not before the ground had swallowed a number of wagons. The fact that these technical details were discussed *en détail* before the Reichstag illustrates a remarkable public interest.[9]

While the area around the Kudensee had the most difficult terrain for building the line of the canal, the locks close to the river Eider on the Western

Fig. 3 Upheaval of earth alongside the Kiel Canal embankments.

of the Panama Canal, London 2007 (don't be repelled by the sensationalist title, part two "The French Tragedy" gives a very good account of technical and other difficulties).

8 For a detailed description of the canal's construction see: Fülscher, Johann, "Der Bau des Nord-Ostsee-Kanal", in: *Zeitschrift für Bauwesen*, vol. 46, 1896, 217-258, 381-408, 495-532; vol. 47, 1897, 117-142, 275-302, 405-452, 525-486; vol. 48, 1898, 41-82, 205-282, 441-490, 693-752; vol. 49, 1899, 99-126, 269-304, 425-464, 621-674, here esp. vol. 46, 1896, 226-227, 501-503, here: Fülscher, "Der Bau des Kaiser-Wilhelm-Canals", 1896, 226-7.

9 See Graf von Holstein's parliamentarian presentation (fn. 2).

terminus posed the biggest engineering challenge of the whole enterprise. Here two basins needed to be build, each 150 meters longs, 25 meters broad, and in order to allow a depth of 8.7 meters during low-tides, the dig needed to reach a depth of approximately 15 meters [fig. 4]. The ground behind the dykes of the river Elbe, themselves not yet a century old, consisted of old river marsh and was heavily laced with streams of groundwater. For a couple of meters, workers and a steam dredger made good progress. But then groundwater started to penetrate the site. For more than one year bigger and bigger shafts were drilled into the ground between the construction site and the river and equipped with ever more powerful pumps to stop groundwater from reaching the ditch. These efforts were unsuccessful and brought no significant lowering of the water levels. At the end of 1891 a floating dredger was brought in and started to slowly chew into the ground.[10]

Because the bottom of the sluice needed to be realised while water was in the pit, the engineers decided to increase the concrete-floor to have a height of 4.2 meters. The original plans were designed with approximately 2 meters, but the lesser quality in underwater-construction made changes necessary. These changes were costly. For one part the excavations needed to be deeper. For the other, much more building material was necessary. In order to have enough concrete at hand, a small industrial complex was build. Between August and December 1892 67.000 cubic meters cement were thrown into the water. When the lock was pumped dry, the bricklaying began, with the material being produced by specially installed workshops nearby.

On 11 September 1894 the walls were completed and the gates installed. Now workers broke the dams down that used to isolate the gates from the river in the West and the water filled canal in the East. On 27 October the first ship passed from the Elbe through the lock into the canal and back again. Workers, engineers, contractors and dignitaries gathered and watched the procession of the steamer while a band played "Deutschland, Deutschland über alles."[11]

Fig. 4 Construction pit of the locks in the East, 1891.

10 Fülscher, "Der Bau", 1896 (fn. 8), 389-395.

11 Fülscher, "Der Bau", 1897 (fn. 8), 449.

2. UNDERSTANDING THE MACHINES OF CHANGE

The fact that a band played Germany's later national anthem was not the only instance that illustrates that there was more under construction than the material world. The locks were important culmination points to celebrate the accomplishments of German technology. At the world exhibition 1893 in Chicago, models of these locks represented Germany and its technological progress at a prominent part in the national pavilion. Here, every lock measured three meters in length. Although no music was playing, the visitors could still watch ship models pass through the lock between the mockup Elbe and canal.[12]

In 1893 Otto Felsing published the sweet and moralistic novel *Im Sturmesbrausen*. Among other things, the book is a showcase how construction work can become a metaphor for the construction of a nation. Seeing the landscape "wrapped in the breeze of the closing dusk, in which man's hands and gigantic machines" worked, his main character perceived the cut as "a horrible open wound". But this does not bother him because it would "someday [be, ECH] a way of honour, of glory and of wealth for the united fatherland."[13] One of the engineers in Felsing's story says that it is "glorious" to build the canal and that he thanks god that he allowed him "to crown the work of my long life with the collaboration at this national work, this national deed!"[14]

The national frame was not the only way to understand the engineering achievements. When one author wrote a book about the "meanings" of the Kiel canal, he noted that it looks "like a miracle [...] when a worker without effort, without trouble, moves the broad, gigantic iron gates [...] like a toy, opens and closes them."[15] Gerhard Stein wrote not about the model in Chicago, but about the 50 times bigger original. The power of technology and the swift control it gives man over nature are in his focus, where the known measures and powers of the human body are transgressed. In the light of this breaking down of the ordinary he speaks of a "miracle" that is defined as a "surprising and welcome event that is not explicable by natural or scientific laws and is therefore considered to be the work of a divine agency."[16] Stein continued in this fashion, obviously awe-struck with the dimensions of the lock and what they represented. The sluice was

> not just great, but beautiful and in every aspect monumental [...] [T]he imposing construction rises, an overwhelming product of the creative human spirit and of human energy. There well exist but a few structures, which in character of execution resemble [... this, ECH]. The ensemble is a true marvel of modern technology.

12 *Der Nord-Ostsee-Kanal, Die Ostseeschleusen zu Holtenau, Begleitschrift zu dem auf der Weltausstellung in Chicago ausgestellten Modell der Schleusen*, Berlin 1893, 4.

13 Felsing, Otto, *Im Sturmesbrausen, Ein Künstler-, Liebes-, u. Streik-Roman v. Nord-Ostsee-Kanal*, Berlin 1893, 69.

14 Ibid.

15 Stein, Gerhard, *Der Nord-Ostsee-Kanal und seine Bedeutung. Eine Schilderung des Nord-Ostsee-Kanals und seiner landschaftlichen Umgebung*, Berlin 1895, 17.

16 "Miracle" in: *New Oxford American Dictionary*, Source: www.oxforddictionaries.com (accessed 12-5-13).

While representing national engineering achievements, the canal for Stein also became a hallmark for the general process of human progress. Again, he chose to understand technological achievements through the realm of the supernatural. He was not alone in referring to such lofty images to express his puzzlement. After he stood on the base of the still dry lock, the ambassador of the Hanseatic towns in Berlin wrote to the senate of Lübeck: The view was "wonderful" and "similar to the camera of a church"[17] [fig. 5].

Under construction was an understanding of progress. The nation as well as human civilization in general were part of the framework to grasp the implications of a technological change that tore the landscape apart. Often the contemporaries solely found the meaning of technology within the paranormal, within the wondrous and the "gigantic" of the machines. The illustrated magazine *Gartenlaube* noted that a steam-dredger is "a curious, but essentially simple monster", a "brute, progressing slowly on its rails".[18] In his novel, Felsing shared these monstrous perceptions of the machines. When his characters travel on one of the construction-railroads, the locomotive turns into an "enormous being, nurtured by fire and brought into being by man's hand", a "huge, chasing and iron animal with steam-snorting nostrils".[19] The notion of the monstrous can also be found in the depiction of machines, for example when he describes "iron steam dredgers, equipped with buckets in front and their sharp steel knifes flashing over the heath"[20].

Size, power and beauty of technology on one hand and the swiftness of control man has over it on the other, jointly constituted a thing for the contem-

Fig.5 Lock on the eastern terminus during construction, 1895. Note the visitors touring the lock.

17 Letter of the Hanseatic Ambassador to Berlin Krüger to Lübeck's senator Klügmann (Stadtarchiv Lübeck NSA 10541, 1885-1896, Herstellung des Nord-Ostseekanals).

18 *Die Gartenlaube, Illustrirtes Familienblatt*, 1889, 846.

19 Felsing, *Im Sturmesbrausen* (fn. 13), 132.

20 Ibid., 43.

poraries to marvel over. The fact that technology gave the worker a supernatural power to handle the gates like toys changed the familiar ideas of size and the possibility of human bodies.[21] Technology was understood through animalistic metaphors that turned dredgers into living creatures. These hybrid beings transgressed the categories of nature and technology and thus contested a dichotomy. Visiting the widening of the canal in 1911, the later World War I general von Moltke noted the amazement that overcame him when he looked at

> the gigantic movement of earth undertaken here. The most heterogeneous machines work everywhere. Here digs an earth-grubber the ground; like a sensible being it bites with its iron jaws into the ground, eats mud, sand, stones, even whole boulders down into itself, rises the earth-filled muzzle and spits after a swift rotation the entire content into a waiting lorry, filling it with one gulp. Then he focusses on the ground again, tearing the mouth open and eating again.[22]

While engineers, workers and machines constructed the Kiel Canal, the contemporaries were fascinated by technology and tried to make sense of it. Fascination means in its Latin origin a bewitching, in this case views that caught the contemporary's attention and that disregarded easy understanding. Some sense of the observed was made through the idea of national and human progress. Reinhart Koselleck understands progress as rationalisations of the accelerated historical change since around 1800.[23] The concept of progress thus closes a gap that was opened between experience and expectation.

Human control over the inanimate in yet unknown magnitudes through technological artefacts is part of this story. The fascinated "bewitched" staring at these artefacts was produced by the sublimity they held for the contemporaries. Like Edward Burke, Martina Mittag identifies the sublime as a secret that defies unravelling and that produces terror as well as wonder.[24] David Nye similarly notes that the "sublime taps into [...] fundamental hopes and fears."[25] Both the monstrous and the marvellous characterise the observation of technological artefacts on the construction site of the Kiel Canal. The secret of this ensemble is the breakup of the dichotomy of nature and technologies through machines that defy these categories. Or, as Minsoo Kang puts it, the sublime originates where the categories "of the animate and the inanimate, the natural and the artificial" break down.[26] While the canal was constructed, the contemporaries struggled to build an understanding of technology as well as the terror and wonder it held.

21 Stewart, Susan, *On Longing. Narratives of the Miniature, the Gigantic, the Souvenir, the Collection*, Baltimore 1984.

22 Moltke, Helmuth von (der Jüngere), *Erinnerungen, Briefe, Dokumente, Ein Bild vom Kriegsausbruch, erste Kriegsführung und Persönlichkeit des ersten militärischen Führer des Krieges*, ed. and commented by Eliza von Moltke, Stuttgart 1922, 360.

23 Koselleck, Reinhart, *Vergangene Zukunft. Zur Semantik geschtlicher Zeiten*, Frankfurt a.M. 1979, 59.

24 Burke, Edmund, *On the Sublime and Beautiful*, Vol. XXIV, Part 2. The Harvard Classics. New York 1909–14; www.bartleby.com/24/2/. (accessed 8-12-12); Mittag, Martina, "Subliming the secret. A genealogical approach", in: Collier, Gordon (ed.), *Critical interfaces: Contributions on Philosophy, Literature and Culture in Honour of Herbert Grabes*, Trier 2001, 99-115, 101.

25 Nye, David E., *American Technological Sublime*, Cambridge (Mass.)/London 1999, xiii.

26 Kang, Minsoo, *Sublime Dreams of Living Machines. The Automaton in the European Imagination*, Cambridge et. al. 2011, 7.

3. HIERARCHY AND HARMONY UNDER CONSTRUCTION

The public reserved even more attention for social policy. The earliest calls for social welfare on the construction site came from the Church. Early in 1886, soon after the state had decided to build the Kiel Canal and before the mass of workers found work along the 100 kilometre long line, Friedrich von Bodelschwingh (1831-1910) gave a presentation. He was the most prominent representative of Protestantism and engaged in preaching inside Germany as well as promoting social welfare for the working class. When speaking in front of an assembly of the Home Mission ("Innere Mission") in Kiel, he adopted the theme of "honour, glory and wealth of the father land" to promote his idea for social policy. He asked his audience to imagine what would happen if the first proud warship with flags flying high and under great festivities crossed from the North Sea to the Baltic and the canal "is on both sides flanked by graves, by the poor victims of alcoholic spirits and the air is filled by the mouldy smell and pestilence of morally decayed people"[27]. To prevent this horrid vision, he outlined a social policy.

For one part the workers had to be saved from their own vices. Alcohol, especially hard spirits, were to be regulated or completely forbidden. The workers should be encouraged or even forced to safe most of their income. Church services also were to be offered regularly. For the other part, the workers had to be guarded against the selfish interest of the contractors. Bodelschwingh despised and feared the spread of the Truck system to the canal's construction site. This system, common in nineteenth century Europe, drove navvies into financial dependency. Contractors would force their employees to sleep and eat at their own barracks and shops, often for exorbitant prices. The workers would be allowed to buy many items but especially alcohol on loan and were left with a faint and unrealistic hope to ever settle their debts. Accordingly, they were forced then to continuously work for their employer. To prevent such extremes, Bodelschwingh envisaged the state to run the barracks and kitchens at the lowest possible self-sustaining price. Overall, he formulated the vision of mediation between the antagonistic forces of irrational behaviour and socialism on one hand and capitalism on the other. On the construction site of the Kiel Canal a vision of German society was under construction.

In his speech, Bodelschwingh designed the blueprints for policy on the Kiel Canal, where state run barracks were put up along the line of the waterway. It was compulsory for the unmarried workers to live and eat there. Members of the social democratic party were excluded from the workforce. To ensure everybody stuck to the rules, the contractors had to take the money out of their worker's wages and pass it straight to the commission running the camps. For these compulsory charges, workers had not only adequate food and a bed, but also regularly cleaned sheets, were able to wash their clothes and attend for their personal hygiene. Church services were held on

27 Bodelschwingh, Friedrich von, *Die Fürsorge für das leibliche und geistliche Wohl der Arbeiter beim Bau des Nord-Ostseekanals besonders auch bezüglich ihres Zu- und Abzugs, Referat gehalten auf dem 11. Jahrestreffen des Landesvereins für Innere Mission in Schleswig-Holstein zu Neumünster am 15. September 1886*, Neumünster 1886, 9.

Sundays. In protestant Schleswig-Holstein, catholic priest who could speak Polish cared for the migrant workers form Prussia's Eastern provinces, Russia and the Danube Monarchy. At the end of 1889, out of 4.600 workers more than 3.000 were living in these quarters.[28]

The press, with the exception of the socialist newspapers, praised the system of the state run barracks. The illustrated magazine "Die Gartenlaube" brought picturesque illustrations that generate an almost cosy atmosphere. The pictures were accompanied by a text of the same general character, for example when it comes to the meals:

> A big, high, spacious room serves for the collective meals. On long, clean tables and benches the whole lot sits down at noon. The food was cooked in three enormous kettles with steam in the clean kitchen. On that day there was rice with potatoes and beef braised together. What was poured into the clean white enameled pots measuring one and a half liter, looked excellent and smelled highly promising.[29]

The newspaper *National-Zeitung* agreed that a "piece of practical social policy for the workers has been realised", and that the arrangements are well a blueprint for any other enterprises.[30] Noticeable was also the good influence the whole arrangement had on the workers themselves [fig. 6]: "With satisfaction observing eyes notice how truly shabby arriving journeymen gradually appear in good and productive gear."[31] A satisfied Kaiser concluded at the inauguration in 1895: "What could be done for the workers was done in accordance with the principles of the Reich's humane social policy."[32] The Reichs-

Fig. 6 The bourgeois public was imagining such clean and sober workers.

28 Heine, *Vom großen Graben* (fn. 1), 97-106.

29 *Die Gartenlaube, Illustrirtes Familienblatt*, 1889, 846.

30 *National-Zeitung*, 12 April 1890.

31 *Die Gartenlaube, Illustrirtes Familienblatt*, 1889, 846.

32 Wilhelm's address at the canals inauguration ceremony in Holtenau on the evening of 21 Juni 1895, cited after *Archiv für Post- und Telegraphie*, July 1895.

tag send a couple of excursions to the construction site that unanimously praised the social policy. The representatives again saw the social arrangements as a "blueprint for other installations". But the most striking description was delivered by a member of the moderate conservative and catholic Zentrums-party, Joseph Lingens (1818-1902): "As a result of these practical ideas, best harmony characterises the relation between the contractors and the workers."[33] The idea behind these praises was that a strong social policy would be a middle way between an unacceptable socialism and a ruthless capitalism. Or, like Lingens phrased it: the construction site of the Kiel Canal holds the chance to truly achieve German unity, a unity that was until then was only realised politically in 1871.

In general, the social policy followed the conservative ideal that strove for stability and protection of basic needs, but, as the historian Thomas Nipperdey puts it, that also had the tendency to prevent democratisation and emancipation of the working classes.[34] This idea of the "patriarchal state" clearly ruled the social ideals that were envisaged on the construction site of the Kiel Canal. Bodelschwingh's rhetoric illustrates this, he wanted to see the social policy to proof that "thorough discipline can help thoroughly". Having received this help, he believed that the workers could have a "small earthly existence", and be content with that. Accordingly, he does not address the workers, like many of his contemporaries he talks about them. While enthusiastically praising accommodation, hygiene, food and the general working conditions at the Reichstag, the conservative representative von Holstein nevertheless admits that he only had a "short time" and therefore had not "to many conversations" with the workers.[35] Otto Felsing's novel, where he captures the general mood towards technology, also represents the conservative imagining of the workers. Complaining about a strike, an engineer remarks about the worker's lack of gratefulness: "Payment is good, food and housing is good, treatment is almost too good."[36] And when a contractor is faced with the workers' demands in a strike, he points at the necessity to protect them from themselves: "Would like to give them more freedom if they'd only know what to do with it."[37]

The hierarchic concept of the patriarchal state was on the construction site of the Kiel Canal realised down to the layout of the barracks [fig. 7]. The camp, as Foucault puts it, is the spatial order of a power that enforces itself by a general visibility. The basic principle of this power is the principle of spacious convolution of hierarchy and surveillance.[38] The camps consisted of barracks which surrounded a two story building. Downstairs was the canteen, while the camp's principal resided on the upper floor.

33 Lingens address to the Reichstag, 14 February 1893, www.reichstagsprotokolle.de/Blatt3_k8_bsb00018681_00526.html (accessed 10-1-12).

34 Nipperdey, Thomas, *Machtstaat vor der Demokratie. 1866-1918*, München 1993, 471.

35 Graf von Holstein's address to the Reichstag, 20 November 1889, http://www.reichstagsprotokolle.de/Blatt3_k7_bsb00018661_00450.html (accessed 10-1-12).

36 Felsing, *Sturmesbrausen* (fn 13), 157.

37 Ibid., 124.

38 Foucault, Michel, *Überwachen und Strafen. Die Geburt des Gefängnisses*, Frankfurt a. M. 1994, 222.

Fig. 7 Idyllic vision of the worker's barracks.

These headmen were mainly retired soldiers and their social superiority was reflected by the elevated status of their quarters from where they symbolically and in reality overlooked the camp. Also, the canteen for the engineers was separated from the worker's, again realising social order in the spatial configuration of the ensemble. High above the camps ruled the clock tower, with its clock hands and bells announcing the order of work, meals and leisure time, but now irrespective of the individual's status. The whole installation was surrounded by a fence and was only accessible through a single gate which was controlled day and night by a watchman.[39] With its kitchen, washing rooms and its laundry the whole ensemble realised a good hygiene and healthy accommodation but also the patriarchal vision. Not only was a canal under construction, also a conservative social utopia.

There are different ways to interpret such policy in an international context. Looking at the construction of the Panama Canal, Tom Frank Peters argues that the sick and death workers in the tropical climate made centrally organised housing necessary.[40] For him this type of organisational thought has its origin in military tradition that was a central element of construction on the nineteenth century.[41] Though the argument is probably not false, a look at railroad construction in Sweden's north illustrates that extreme climates do not necessarily lead to a centralised organisation of food and housing.[42] One also has to reflect that social policy, as Alexander Missal notes, shows the convictions of US-elites and materializes the imagined differences of race and class in the layout of the barracks as well. [43]

39 "Grundsätze betreffend die Annahme von Arbeitern zum Bau des Nord-Ostsee-Kanals, den mit denselben abzuschließenden Arbeitsvertrag, ihre Unterbringung und Verpflegung, die geistliche Versorgung derselben und die für sie zu treffenden Wohlfahrts-Einrichtungen", November 1885 (Geheimes Staatsarchiv preußischer Kulturbesitz I. HA Rep 84a Justiz-Ministerium Nr. 4866); Beseke, Carl, *Der Nord-Ostsee-Kanal, Entstehungsgeschichte, Bau, Bedeutung*, Kiel 1893, 48.

40 Peters,Tom Frank, *Building the nineteenth century*, Cambridge (Mass.)/London 1996, 316.

41 Ibid., e.g. 356.

42 Theander, Agge, *I rallarnas spår, Tracing the Navvies*, Narvik 1993; a critical view from one of the people on the construction site offers Rosén, Gustav, *Förhållanden och Misförhållanden vid Ofotenbanan*, Umeå 1902.

43 Missal, Alexander, *Seaway to the future, American social Visions and the Construction of the Panama Canal*, Madison 2008. See also Patel, Kiran Klaus, *"Soldaten der Arbeit". Arbeitsdienste in Deutschland und*

Although it is difficult to let marginalised classes speak in historiography,[44] a different story of the Kiel Canal needs be told from the perspective of the workers. Such a narrative is possible, besides a general lack of sources, because of one publication. After he had preached on a number of construction sites of canals, reverend Jakob Küßner published a book with many impressions from the Kiel Canal. What makes him so different from his contemporaries is that he talked to the workers and was engaged in their problems. Very interesting is the worker's completely different evaluation of the barracks. The navvies felt "hate and disgust" for them. Even when offered good barracks like at the Kiel Canal, they always preferred private accommodation, even if it was of the most primitive sort. Where they were forced to move to the barracks, like at the Kiel Canal, "they do so without exception grudgingly and unwillingly". Even with the long working hours they prefer a private room one hour away from any camp.[45]

Life in the camps is too "herd-like", Küßner notes, and where so many people are on one spot, drinking seemed unavoidable. Also the lack of women, with the exception of prostitutes, and family life produced a "raggedness with a certain unavoidability where so many of the same sex are crowded in a small space." Küßner continues arguing that it was misleading to compare the situation with life in the military. Army barracks are often in bigger cities with a variety of entertainment. Also every recruit knows his existence as a soldier was only temporary. Not so the navvies, the clergy man writes. These men had seen all the hardships of life, know that they are stuck with their profession and "have left hope behind". He compares the life of the canal worker with that of the convict, with the restraints of one replaced by the "fear of the highway" and the "promise of a high salary" of the other.[46]

Küßner does not regard workers as his equals, in fact he thinks they are "like the children" because they are often impulsive: "Sometimes you have the wildest blasphemies [...] and often close by you can be the witness of deep emotion and devotion, interjected by sobs, that unsettle yourself."[47] He formulates solutions for the "strict and rational selection" that helps picking the good elements and allows discarding the bad ones. In regard of this solution for the problems unruly workers pose for social order, he scarcely differs from other contemporaries: he demands strict control over the roads around the construction sites and to send away any vagabonds or to fire those workers who are found drunk or behaved unruly in other ways.[48] Thus he, like Bodelschwingh, believed in the thorough help, thorough discipline might bring. Nevertheless, by simply talking with the workers and emphasising with their situation, he produces a narrative that is questioning the public representations of social policy.

den USA 1933-1945, Göttingen 2001, 129ff.

44 See e.g. Ginzburg, Carlo, *The Cheese and the Worms. The Cosmos of a Sixteenth Century Miller*, Baltimore 1980.

45 Küßner, Gustav, *Was sind wir unseren Kanalarbeitern schuldig? Im Auftrag des Deutschen Vereins gegen den Mißbrauch geistiger Getränke*, Leipzig 1905, 2.

46 Ibid., 2-4.

47 Ibid., 45.

48 Ibid., 11.

4. CONCLUSION

On the construction site of the Kiel Canal, engineers, workers and machines dug a long and deep ditch through the province Schleswig-Holstein. While they changed a defiant material world with great difficulty, the public imagination of the construction site was occupied with the artefacts used and installed as well as with the social policy towards the navvies. While the machinery inspired sublime awe and was interpreted as a symbol of national and human progress, social policy seemed to hold the key for a harmonic but hierarchic society.

The construction site of the Kiel Canal was a place where social power and order was realised in discourse, in social policy and down to the layout of the worker's barracks. But the bourgeois utopia was not shared by the workers. Caught up in patriarchal ideas of the feudal society, the conservative view ignored the fundamental democratic changes of the age. Yet, it was also a place where the other side of power is revealed, hate and disgust for the barracks and an outlook on life that reminds of the existence of a convict.

To take the materiality of infrastructure construction serious requires us to consider the environmental impacts of building activity. Especially as the Kiel Canal was a monumental intervention into the landscape of Schleswig-Holstein. The 120 kilometre long stretch of the river between Rendsburg and the mouth to the North Sea was cut off from its upper section. The engineers were surprised that the construction of the Kiel Canal had consequences for the lower Eider, as their careful considerations were proven wrong when sedimentation rapidly filled the riverbed. The construction of the Kiel Canal was the main reason for a developing environmental crisis along the Eider.[49] Fishing decreased, saltwater pushed high up river and deteriorated farming and husbandry, but especially unprecedented storm floods devastated the countryside [fig. 8].

Fig. 8 Photograph from the Eider flood of autumn 1935.

49 For a detailed story of the ecological impacts see Heine, Eike-Christian, "Two canals, two barrages and the remnants of a river. Nature and technology along the Eider, Schleswig-Holstein's longest river", *Environment and History*, accepted but not yet published, for a preview see http://www.whp-journals.co.uk/EH/EHpapers.html (accessed 15-7-15).

Throughout several years of increasing water levels, overspill and destruction, the region debated technological solutions in order to find protection from the unruly river. In 1936 a floodgate was constructed in the middle of the lower river. While the land behind these gates was protected, the situation worsened in the part of the river that was still open to the forces of the North Sea. After the catastrophic storm flood of 1962, another monumental project started to finally resolve the ecological crisis the Kiel Canal had brought. Since 1973 the mouth of the Eider is cut off from the North Sea by more than 4 kilometres of dykes and a large flood gate. The river now is subjected to technological control [fig. 9]. To consider the materiality of construction sites also means to account for the stories where the discursive finds its limitations. The defiant swamps that put engineering skill to a test or the unexpected environmental consequences of construction work show that some things "are first and foremost themselves, despite the many meanings we find in them", as William Cronon notes about objects and environmental history: "We may move them around and impose our design upon them. We may do our best to make them bend to our wills. But in the end they remain inscrutable"[50]. Trying to include their elusive stories into our narratives about the worlds, people build with their heads and hands, will help us to write a better and more insightful history.

Thinking about Cultural History ("Kulturgeschichte"), Ute Daniel demands to reflect, what history means for us personally as well as our age.[51] When working on the history of infrastructure as well as construction sites, I believe we need to reflect on two things. First, we need to have the question of an environmental history in mind that takes the broader relationships between historical actors, their living environment and their

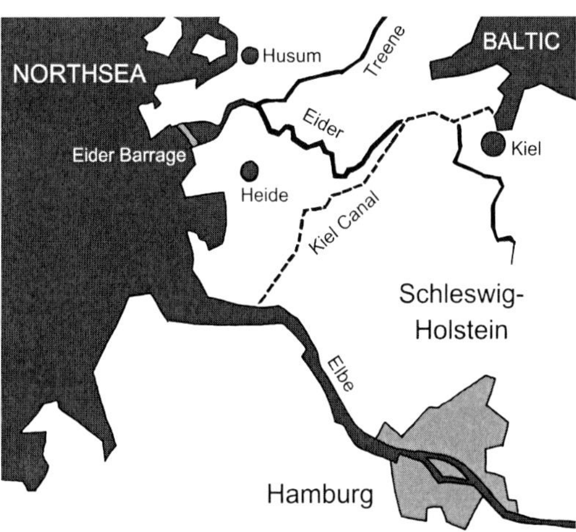

Fig. 9 The map shows how the Kiel Canal cut through the Eider and indicates the location of the barrage at the river's mouth.

50 Cronon, William, "Introduction. In Search of Nature", in: Ibid. (ed.), *Uncommon Ground. Rethinking the Human Place in Nature*, New York/London 1995, 23-56, 55-6.

51 Daniel, Ute, *Kompendium Kulturgeschichte. Theorien, Praxis, Schlüsselwörter*, Frankfurt a.M. [4]2004, 19.

technological artefacts into account. Second, we need to be aware of power and hierarchy between classes (or gender) that were produced or contested by infrastructures. To me this leads to two implications. We must not fall for the sublime horror or bliss of technology, but must confront supposedly practical constraints and insist on democratic control over it.[52] Also, the seemingly self-exposing rhetoric of nineteenth century social-political discourse should make us aware about the modes, in which our time imagines and justifies inequality.

LIST OF FIGURES

Fig 1: Bulius, *Eine Fahrt durch den Kaiser-Wilhelm-Kanals, Die neue Wasserstraße zwischen der Nord- und Ostsee*, Breslau 1895, inside of the cover.

Fig. 2 & 3: Loewe, Carl, *Geschichte des Nord-Ostsee-Kanals, Festschrift zu seiner Eröffnung am 20./21. Juni 1895*, Berlin 1895, n.p.

Fig. 4: Library of Congress, http://hdl.loc.gov/loc.pnp/cph.3a39671 (accessed 6-1-15).

Fig. 5: *Die Gartenlaube – Illustriertes Familienblatt 1895*, 396.

Fig. 6: Loewe, Carl, *Geschichte des Nord-Ostsee-Kanals, Festschrift zu seiner Eröffnung am 20./21. Juni 1895*, Berlin 1895, n.p.

Fig. 7: *Die Gartenlaube – Illustriertes Familienblatt 1889*, 844.

Fig. 8: *Wochenblatt Landbauernschaft Schleswig- Holstein*, 22 February 1936.

Fig. 9: Map drawn by Thomas Schuetz and used with his permission.

52 Nye, David E., *Technology Matters, Questions to live with*, Cambridge et. al. 2007, 226.

PART 6: CONCEPTUALISING CONSTRUCTION SITES

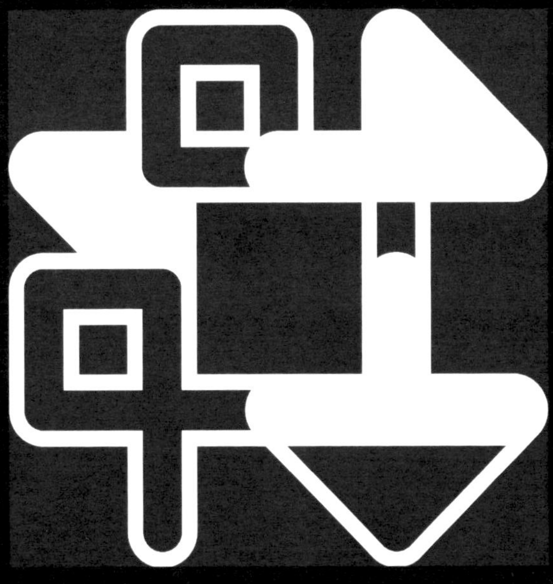

WHAT TO LOOK AT?

NOTES ON THE HISTORIOGRAPHY OF NINETEENTH CENTURY CONSTRUCTION SITES

Christoph Rauhut

On researching construction sites and their history, a number of challenges become apparent:[1] Amongst others is the problem of working comparatively, since each construction site is distinctively different to others and since each site is always a short-lived and continuously changing production facility. The subsequent challenge is to find description concepts that accurately portray the changes and differences between different construction sites in a discursive and abstract manner. One way to summarize and also to judge the processes that characterise construction sites is the use of terms that imply certain processes of transformation: terms like rationalisation, mechanisation, professionalization, standardisation or economisation. Using these terms one can cogently describe the actions and configurations on a construction site (also in a negative manner) and also argue comparatively by using the greater narrative as a contrast against a specific example.[2] It is crucial, however, to note that these terms represent concepts of transformation, which are not solely linked to the realm of building, rather, they are broader terms that originate from modernity's belief that the technical culture is a culture of progress. They incorporate definitions and viewpoints that are strongly linked to an ideology of progress. Thus, they ultimately also pre-set interpretations and results. And to add to the complexity, their use and definition has constantly changed and still changes both in everyday use and in historiography.

This paper focuses on the complex interplay of such concepts in the context of construction sites and thus addresses a key question when discussing the relation between their physical and imaginary landscape. It asks for the specific roles of the physical and the imaginary, when establishing our understanding of construction sites and writing their history.

The aim of such an endeavour is to argue two theses:

The first is the necessary re-assessment of these concepts in regards to the realm of construction. Within the history of technology,

1 This paper uses results from the author's dissertation entitled *Die Praxis der Baustelle um 1900: Das Zürcher Stadthaus Fraumünsteramt*. Dissertation Zürich 2014. Further notable works on constructions sites for the nineteenth and early twentieth centuries are: Schumacher, Thomas, *Großbaustelle Kölner Dom, Technik des 19. Jahrhunderts bei der Vollendung einer gotischen Kathedrale* (*Studien zum Kölner Dom*, 4), Köln 1993; Bertels, Inge, *Building the city, Antwerp 1819-1880*, Dissertation Leuven 2008. Valuable for Medieval Ages is: Binding, Günther, *Baubetrieb im Mittelalter*, Darmstadt 1993.

2 E.g.: Peters, Tom Frank, *Building the nineteenth century*, Cambridge (Mass.)/London 1996; Renz, Kerstin, *Industriearchitektur im frühen 20. Jahrhundert. Das Büro von Philipp Jakob Manz*, München 2005. Further references will be given in the following in respect to specific terms.

concepts such as rationalisation, mechanisation, professionalization, standardisation, economisation, etc. have an established tradition of use: They have been discussed and disputed, located adjacent to other concepts, and applied over and over again. Although they have not yet been historicized themselves, their use and definition have changed over time and still changes today.[3] One can say that they are rather well established. However, when trying to use these concepts in the realm of construction it becomes apparent that they are somewhat inadequate or even misleading. Most concepts remain focused on what can be called their classical fields of interest, for instance, the history of technology being primarily focused on industrial production [fig. 1]. If we want to use these concepts as descriptive tools detailing the activities of a construction site [fig. 2], then they need to be reasserted or redefined in a specific way in relation to the field. Such a necessity also illustrates the link of the endeavour to the field of *Begriffsgeschichte*.[4] Here, Reinhart Koselleck named a central argument that a history of concepts necessarily implies a certain level of ambiguity,[5] something that is also the basis to redefine them within the realm of construction.

The second thesis is that we need to be more aware of the construction sites we choose to consider in our research. To paraphrase a second central concept of Koselleck's notion of *Begriffsgeschichte*:[6] We have to pay attention to the particular historical circumstances. It seems that when writing the history of the construction site, at least for the nineteenth and early twentieth centuries, one tends to look at the

Fig. 1 AEG production shed for alternating current motors in Berlin, photo 1912.
Fig. 2 Construction site of the "Gull-Haus" in Zürich, 1886.

3 Examples of the recent discussion are David E. Nye's book on manufacturing by assembly line (Nye, David E., *America's Assembly Line*, Cambridge (Mass.) 2013) or the 2014 annual conference of the Gesellschaft für Technikgeschichte in Stuttgart, which had the topic: "Produzieren, Herstellen, Fabrizieren: Neue historische Perspektiven auf die Produktionstechnik".

4 Most influential is maybe the work of Reinhart Koselleck, who stressed the concept of *Begriffsgeschichte* and its possibilities and restraints from various viewpoints As a compilation of the most important texts by Koselleck: Koselleck, Reinhart, *Begriffsgeschichten*, Frankfurt a.M. 2006. Summarising the recent methodological discussion in the field: Eggers, Michael/Rothe, Matthias, "Die Begriffsgeschichte ist tot, es lebe die Begriffsgeschichte!", in: ibid. (ed.), *Wissenschaftsgeschichte als Begriffsgeschichte*, Bielefeld 2009, 7-22, here 7-14.

5 Stressing this point: Bud, Robert, "Introduction" [for the issue: "Focus: Applied Science"], in: *Isis*, vol. 103, 2012, Nr. 103, 515-517, here 516.

6 Ibid.

construction sites that are already well known and regarded as symbols of certain concepts at their time.[7] The Paris Panthéon (1764-1790) was always seen as a milestone for modern construction techniques, the Crystal Palace (1851) remains the origin of construction as assembly [fig. 3], to name two vivid examples. In fact, such association can also be observed in today's discussion of construction sites, only the disputed terms have changed.[8] The thesis is that this choice, and this system of choice, needs to be re-evaluated if one would like to write a more complete history of the construction site and its underlying concepts. Such a history needs to take into account the broader building activity and all the 'physical' changes and not only those that defined the imagery of construction sites at the time.

The aim of the following is to verify these two theses by considering the two concepts rationalisation and mechanisation with special regard to construction sites in detail. The claim is, nevertheless, that such a discussion could also be made in regards to any other transformation concept that has been used by modernity to describe the changes of and within technical culture.

Fig. 3 Construction site of the Crystal Palace for the First London International Exposition of 1851.

7 The focus is generally on the large construction sites, e.g.: Peters, *Building the nineteenth century* (fn. 2); Grisby, Darcy Grimaldo, *Colossal. Engineering the Suez Canal, Statue of Liberty, Eiffel Tower, and Panama Canal*, New York 2012; Pfammatter, Ulrich, *In die Zukunft gebaut. Bautechnik und Kulturgeschichte von der Industriellen Revolution bis heute*, München/Berlin/London/New York 2005.

8 A vivid example is our definition of a *Grossbaustelle*. See e.g.: Böhme, Hartmut, "Baustelle ist überall", in: *Neue Zürcher Zeitung*, 8 December 2012. (http://www.nzz.ch/aktuell/feuilleton/literatur-und-kunst/baustelle-ist-ueberall-1.17876478, 25 February 2013).

1. RATIONALISATION

The concept of rationalisation has been intensively discussed for many decades, within different contexts, historiographical traditions, and interpretations.[9] In the context of this paper, one might focus on important aspects of the German concept of rationalisation, a concept that is mutually dependent on the English and American equivalents, but which is at the same time, as Akos Paulinyi said, essentially a German creation.[10] One basal aspect is that, until the 1980s, the concept of rationalisation had mostly been discussed as relating to industrial production. It was used to describe how the nature of work and the organisation of labour were transformed by and within factories. Only recently new perspectives have been discussed that use the concept of rationalisation to describe beyond the division and optimization of labour, while continuing to focus on efficiency and placing importance on consumer choice[11] or the role of trust.[12] Moreover, the use of the concept was broadened to apply to all fields of production and work. Recent publications such as *Rationalisierung in Handwerksberufen* reflect this trend.[13] Rationalisation was even defined as common, reasonable human behaviour.[14]

However, the historiography of construction and construction sites so far ignores these conceptual innovations. Instead, two narratives are still dominant.

The first one is the long-held view that the average construction site in the nineteenth and early twentieth centuries reflected the opposite of a rationalised vision. This view seems to originate from the influential works of the Verein für Socialpolitik in the late nineteenth century.[15] Alongside other studies, this institution published studies on the building trades in particular cities, developing on the idea that construction cannot be rationalised: "Construction [preserved] its handicraft character due to the extremely limited use of machines as well as due to an inapplicability of a furthered division of labour."[16] This viewpoint is derived by comparing the

9 As a summarisation of the *Rationalisierungsbewegung* and corresponding evaluations: Heßler, Martina, *Kulturgeschichte der Technik* (*Historische Einführungen*, 10), Frankfurt a.M./New York, 38-71; Reith, Reinold, "Rationalsierung im Handwerk: Ein Widerspruch?", in: LWL-Freilichtmuseum Hagen (ed.), *Rationalisierung in Handwerksberufen* (*Forschungsbeiträge zu Handwerk und Technik*, 24), Hagen 2012, 8-25, here 10-11.

10 Paulinyi, Akos, "Massenproduktion und Rationalisierung", in: *Technikgeschichte*, vol. 56, 1989, 131-181, here 173-174.

11 Heßler, *Kulturgeschichte der Technik* (fn. 9), 38.

12 Frevert, Ute, "Trust as Work", in: Kocka, Jürgen (ed.), *Work in a Modern Society. The German Historical Experience in Comparative Perspective* (*New German Historical Perspectives*, 3), New York 2010, 93-108.

13 LWL-Freilichtmuseum Hagen (ed.), *Rationalisierung in Handwerksberufen* (*Forschungsbeiträge zu Handwerk und Technik*, 24), Hagen 2012.

14 VDI (ed.), *Rationalisierung heute*, Düsseldorf 1988, cited after: Reith, "Rationalsierung im Handwerk: Ein Widerspruch?" (fn. 9), 10: "allgemeines, vernüftiges menschliches Verhalten".

15 Reith also sees this point in view of the entire craft sector. Reith, "Rationalsierung im Handwerk: Ein Widerspruch?" (fn. 9), 14.

16 Kreuzkam, Theodor, "Das Baugewerbe mit besonderer Rücksicht auf Leipzig", in: *Untersuchungen über die Lage des Handwerks in Deutschland mit besonderer Rücksicht auf seine Konkurrenzfähigkeit gegenüber der Großindustrie*, Vol. 9: *Verschiedene Staaten* (*Schriften des Vereins für Socialpolitik*, 70), Leipzig 1897, 543-628, here 548: "das Baugewerbe [bewahrt, CR] seinen handwerksmäßigen Charakter sowohl infolge eines unverhältnismäßig beschränkten Gebrauches der Maschine, als [auch, CR] der Unanwendbarkeit einer weitgehenden Arbeitsteilung und Arbeitsgemeinschaft."

construction site with industrial mass production spaces, such as factories. It is a position ever present in contemporary research: to cite a more recent work on the building sector of Zurich: "The production of buildings, especially for housing, has [comparatively] been a very slow progressing sector of industry until the twentieth century."[17] Until today rationalisation is not seen as a concept that can be applied to the average construction site of the nineteenth and early twentieth century.

The second influential narrative about construction sites argues that rationalisation was only be applied to industrial buildings. Such an equating of construction sites of industrial buildings and the concept of rationalisation was mainly based on the construction materials and products used for industrial buildings. Their standardisation (another term one might critically discuss) was seen as an opportunity to structure the work more efficiently in order to rationalise the sites.

The equating of rationalisation and the sites of industrial buildings was widely accepted at the time, this was the imagery perception, since contractors and engineers strongly propagated such an interpretation. Good examples for such a case are the sites for the industrial buildings designed and executed by Philipp Jakob Manz's company, researched in detail by Kerstin Renz.[18] Manz's company erected numerous industrial buildings in the early twentieth century and they were famous for their speed in the planning and building process. Manz did a great deal to promote the efficiency of his works on all levels; he even called his company 'Blitzbüro Manz'.[19]

This association that was established at the time is influential until today, Kerstin Renz argues for example: "In the construction of industrial buildings, the tendency to streamline construction processes is self-justified by the building type."[20] And Rainer Kaden, who has worked on the mechanisation of construction, agrees: "High performance requirements necessitated rationalising the building production. Part of the respective tasks are [...] the construction of industrial buildings."[21] The concept of

---

17 Bärtschi, Hans-Peter, *Industralisierung, Eisenbahnschlachten und Städtebau. Die Entwicklung Zürcher Industrie- und Arbeiterstadtteils Aussersihl*, Basel/Stuttgart 1983, 253: "Die eigentliche Bauproduktion, insbesondere der Hausbau, [gehört, CR] bis ins 20. Jahrhundert zu den zurückgebliebensten Industriezweigen". A similar viewpoint expresses e.g.: Lüthi, Christian, "Baugewerbe", in: *Historisches Lexikon der Schweiz*, http://www.hls-dhs-dss.ch/textes/d/D41548.php, (accessed 24-11-2009); John, Peter, "Bauhandwerk und Industrie. Von den Gesellenverbänden zur Gewerkschaftsbewegung", in: Reese, Hartmut et al. (ed.), *Hand in Hand. Bauarbeit und Gewerkschaften. Eine Sozialgeschichte*, Frankfurt am Main 1989, 12-27; Plumpe, Werner, "Wirtschaftsgeschichtlicher Überblick. Entwicklung und Struktur des deutschen Baugewerbes", in: Reese, *Hand in Hand* (same fn.), 364-373. Some recent work reflects the trend to judge less harsh, but it still addresses the slow modernisation of work at the average construction site, e.g.: Schmiek, "Modernität in Raten – Aspekte ländlichen Bauens zwischen 1880 und 1930 im nördlichen Oldenburg", in: Dahms, Geerd et al. (ed.), *Stein auf Stein. Ländliches Bauen zwischen 1870 und 1930* (*Arbeiten und Leben auf dem Lande*, 6), Kiekeberg 1999, 171-205.

18 Renz, *Industriearchitektur* (fn. 2). Another example is the company of Boswau & Knauer. Haps, Silke, *Industriebetriebe der Baukunst – Generalunternehmer des frühen 20. Jahrhunderts. Die Firma Boswau & Knauer*, Dissertation Dortmund 2008.

19 Renz, *Industriearchitektur* (fn. 2), 58.

20 Ibid., 54: "Die Tendenz zum rationalisierten Bauprozeß ist beim Industriebau in der Bauaufgabe selbst begründet."

21 Kaden, Rainer, "Einflüsse aus der Meachanisierung von Bauprozessen auf die Entwicklung von Bauverfahren", in: Sebastian, Ursula (ed.), 5. *Kolloquium Geschichte der Bauingnieurwissenschaften*, Leipzig 1989, 41-49, here 19, cited after: Renz, *Industriearchitektur* (fn. 2), 55: "Ausgegangen sind Impulse zur [...] Rationalisierung in der Bauproduktion [...] von hohen Leistungsanforderungen an das Bauwesen.

rationalisation is not linked to further changes in the realm of construction; it is rather the opposite: It is implied that no other forms of rationalisation exist.

However, the activity of the Bauhaus in Dessau-Törten is, as Andreas Schwarting has shown,[22] a prominent example that such consideration does not even always prove true. Dessau-Törten (1926-1928) was one of the larger construction activities at the time and it was important for the Bauhaus and its proponents that it represented the progressiveness of architecture and society. Already the aesthetics were meant to display the *Rationalität*, thus modernity, of the project. One also tried to make such an idea present in the construction process. Photos and texts tried to convince contemporary audiences about the site's progressiveness [fig. 4], achieved by the rationalisation and mechanisation of the work. However, one has to be careful of these contemporary ascriptions. Looking closely at the site and putting the site into their historical context, many of the innovations were actually not used or of less significance than reported. Schwarting titled this the myth of runway cranes.[23] To illustrate it boldly: It was claimed that prefabricated concrete elements were to be used in Dessau-Törten, but many parts were actually built on site in brick and mortar.

Nevertheless, the work of Manz, of the Bauhaus in Dessau-Törten, or similar projects involving stakeholders longing for public visibility, were a propagandistic success, since they were able to stage that rationalisation was only to be achieved on sites of industrial buildings.

Fig. 4 Representation of the construction site of Dessau-Törten in contemporary publishing by the Bauhaus.

Dazu gehören Aufgaben [...] bei der Errichtung von Industriegebäuden und -anlagen."

22 Schwarting, Andreas, *Die Siedlung Dessau-Törten. Rationalität als ästhetisches Programm*, Dresden 2010.

23 Ibid., 12: "der Mythos der Kranbahnen".

However, putting the two described narratives aside and using some of the more recent results from research into rationalisation, one might redefine the meaning of rationalisation in regards to construction sites: One might even express that the average construction site was rationalised in the nineteenth century. There is much evidence for this proposition, even if one looks at sites that have never been considered rationalised, such as those which accounted for the typical average building activity at the time like housing and commerical buildings.

The central argument is the work system that describes the practices on a construction site. The construction site was always a system of different trades with different tasks, thus differentiation of work by types of work is almost a genuine characteristic of construction sites [fig. 5]. Furthermore, by looking closely at the operations of a late nineteenth century construction site, as I have done in my doctoral research with a particular focus on the site of the office building *Fraumünsteramt* in Zurich (1899-1900),[24] one can discern that this systematic approach was strengthened and expanded over the second half of the nineteenth century. New and specialised stakeholders increasingly appeared, such as masons who specialised in specific types of brickwork, glaziers who concentrated on roof windows, workers who were trained in glass tile construction [fig. 6] or special parquet recliners who took over a task that was traditionally assigned to carpenters. Such partakers are hard to discover in retrospect because they were often hired as subcontractors and as such not necessarily included in official documents, but their contributions made up the collective effort of rationalisation.

Another part of the process to rationalise construction was the system of wages. Reinhold Reith raises this issue in his telling work *Lohn*

Fig. 5 Workers at the site of the office building *Fraumünsteramt* in Zurich, 1899.

24 See: Rauhut, *Die Praxis der Baustelle um 1900* (fn. 1). This work includes a detailed listing of the archival sources used.

Fig. 6 Special stakeholder for the erecting of a glass vault on the site of the office building Fraumünsteramt in Zurich, April 1900.

und Leistung.[25] He points out that there is a close interrelation between the work process and the type of wage, the main distinction being between piece wages and time-rate wages. Additionally, Reith shows how piece wages created a more flexible work setting by increasing the elasticity of labour provided.[26] The wage system implemented on the construction site of the office building *Fraumünsteramt* accords with this conclusion. The sculptural work of the stonemasons is an example: Highly skilled tasks were paid at an hourly rate to ensure the required quality, whilst standard work was paid at a piece rate, allowing the number of workers to be quickly increased and also motivating the labourers to work harder. Rationalisation on the construction site also involved the implementation of a more diverse wage system.

One might introduce further cases of rationalisation processes, but the given examples are sufficient to sum up the concept of rationalisation and its (possible) use: Despite opposing prophecies, there is very little doubt that the construction site of the late nineteenth century can be described as a workplace undergoing rationalisation, especially if using the more recent definitions of the concept. However, to be able to do so, it is critical not to focus on construction sites that were said to be rationalised in the historical context, but to reassess our selection, to include the broader building activity of the period.

25 Reith, Reinold, *Lohn und Leistung. Lohnformen im Gewerbe 1450–1900* (*Vierteljahreschrift für Sozial- und Wirtschaftsgeschichte*, Beihefte 151), Stuttgart 1999.

26 Ibid., 17, 21.

2. MECHANISATION

Mechanisation is yet another concept that should be critically reassessed in regards to its application to historical construction sites. Generally, it is a concept frequently used to describe transformations, especially in regards to changes in production processes within the nineteenth century.[27] A discussion of the concept of mechanisation, however, includes some challenges: The first is that the concept of mechanisation interestingly lacks the numerous definitions that one finds, for example, for terms such as rationalisation, especially if considering the German equivalents.[28] One reason for these challenges seems to be the ambiguity of the concept; it has changed over time and differs between different thematic fields.[29] The second is that the English term mechanisation has, simplified, two German equivalents: *Mechanisierung* and *Maschinisierung*.[30]

To glimpse at the historiography of the concept, one may focus on the definitions of these two German terms: nineteenth century use describes *Mechanisierung* as the division of the individual work process steps that can be substituted by machines. The substitution of steps within the production by machines is the aspect of *Maschinisierung*.[31] These definitions were closely linked to industrial production and derived from the changes in production in the 'classic' fields of industrialization in the nineteenth century. More recent definitions differentiate both terms more strongly and broaden the definitions: *Mechanisierung* is described as the support of human labour by the use of machines, aspects included might also be an improved power gear ratio or the addition of tool functionality;[32] *Maschinisierung* is generally linked to the substitution of the human work process by machine production.[33]

In construction history, mechanisation is mostly linked to the use of (larger) building machinery: Bucket excavators, revolving cranes or cement mixers were thus the icons of mechanisation for the second half of the nineteenth century and the beginning of the twentieth century. Construction sites that saw such machinery are labelled mechanised. An early but prominent example of this is the Suez Canal (1859-1869):[34] Here human workers were

27 Most prominent to discern in: Paulyini, Akos/Trotzsch, Ulrich, *Mechanisierung und Maschinisierung. 1600 bis 1840* (*Propyläen Technikgeschichte*, 3), Berlin 1997.

28 A definition of *Mechanisierung* is not included in, for example, the technical encyclopaedias of the early twentieth century (such as Lueger, Otto (ed.), *Lexikon der gesamten Technik und ihrer Hilfswissenschaften*, 8 vols., Stuttgart/Leipzig [2]1904-1920), but also not in recent encyclopaedias (such as: Jäger, Friedrich (ed.), *Enzyklopädie der Neuzeit*, 16 vols., Stuttgart/Weimar 2005-2012).

29 A good example is the field of agricultural works, the term has a rather specific meaning in this context, but which changed over time. As an overview: Bruckmüller, Ernst/Langthaler, Ernst/Redl, Josef (ed.), *Agrargeschichte schreiben. Traditionen und Innovationen im internationalen Vergleich* (*Jahrbuch für Geschichte des ländlichen Raumes*, 1), Innsbruck 2004.

30 As the works cited in the following are mainly German texts, it seems appropriate to focus on the use and definition of the German term. A short introduction on the use and definition of the English term offers: Mechanization, in: http://en.wikipedia.org/wiki/Mechanization (accessed 21-3-2014).

31 Cited definition for the nineteenth century follows: Paulyini/Trotzsch, *Mechanisierung und Maschinisierung* (fn. 27), 146.

32 Voigt, Kai-Info, "Mechanisierung", in: Springer Gabler Verlag (ed.), *Gabler Wirtschaftslexikon*, http://wirtschaftslexikon.gabler.de/Archiv/72567/mechanisierung-v4.html (accessed 21-3-2014).

33 Ibid.

34 Discussing the construction of Suez Canal: Grigsby, *Colossal* (fn. 5), 42-68; Peters, *Building the 19th Century* (fn. 7), 178-204.

substituted by large machinery such as ladder excavators or dredges [fig. 7]. The main reason for this change was political, since there was no longer access to an abundance of cheap labourers. However, within the very context of such a great and stunning endeavour this change in the work process had to be labelled as a sign of technical progress:[35] The substitution of human labour by machinery had to be a result of the process of mechanisation.

Another prominent example of a rather propagandistic use of the term mechanisation is the Centennial Hall in Breslau (today Wroclaw) (1911-1913).[36] This site was incredibly well perceived at the time. Not only because the building under construction featured the largest reinforced concrete dome at the time, but also because the engineering firm advocated the site in various media, such as articles in journals.[37] One aspect that received special attention in the coverage was the cableway crane used [fig. 8]. This machine substituted, in a single

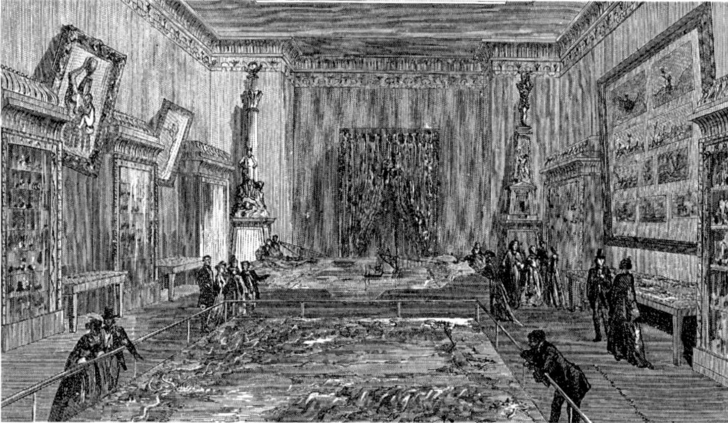

Fig. 7 Room in the Suez Canal Company Pavilion at the International Exposition of 1867 in Paris.
Fig. 8 The cableway crane at the construction site of the Breslau (Wroclaw) Centennial Hall, printed in one of the most important contemporary engineering journals, 1913.

35 On the representation of the site in Europe: Grigsby, *Colossal* (fn. 7), 52-56.

36 Ilkosz, Jerzy, *Die Jahrhunderthalle und das Ausstellungsgelände in Breslau. Das Werk Max Bergs*, München 2006.

37 Stegmann, Knut, *Das Bauunternehmen Dyckerhoff & Widmann. Zu den Anfängen des Betonbaus in Deutschland 1865–1918*, Dissertation Zürich 2010, vol. 1, 253. Another good example is the book: Trauer, Günther/Gehler, Willy, *Die Jahrhunderthalle in Breslau*, Berlin 1914.

step, transportation that previously necessitated a combination of different machines. This change in transportation was often linked to the concept of mechanisation because this was the image deliberately displayed at the time.

Here, construction sites with such highly visible machinery prominently staged their definition of mechanisation and were thus able to define the general concept. As a result, construction sites without such emblematic machines often seemed not to be mechanised. A view still present in contemporary literature: "Instead of working with cranes one used lifting blocks and hoists: bucket excavators were not used. [...] The mechanisation on site is hardly advanced."[38] Such a conclusion is however only based on the concept of mechanisation in the nineteenth century with its focus on large and visible building machinery. Another understanding of the term mechanisation might, in contrast, even be seen as a generally prevalent aspect of construction sites of that era.

A starting point is to look at the transport system for the building materials: During the middle of the nineteenth century a rather distinct and elaborated interrelated system for the vertical and horizontal transport of goods became common.[39] It included new machinery, travelling cranes and local track systems, which were the more visible features [fig. 9], but also integrated small devices such as hoisters or special lifting blocks. Weston's differential pulley block was, for example, a big deal at the time.[40] Mechanisation might in this case be described as the support of human labour by the use of machines, but not its substitution. In fact labour substitution

Fig. 9 Travelling crane on the site of the office building *Fraumünsteramt* in Zurich, 1899.

38 Renz, *Industriearchitektur* (fn. 2), 67: "Man arbeitet anstatt mit Kränen mit Flaschenzügen und Winden, Löffelbagger kommen nicht zur Anwendung. [...] Die Mechanisierung auf der Baustelle ist noch kaum fortgeschritten."

39 A good description of this system offers: Schumacher, *Großbaustelle Kölner Dom* (fn. 1), 545-570, 781.

40 There are numerous articles on this pulley block in the technical journals, e.g.: Anonymous, "Weston's Flaschenzug", in: *Dingler's polytechnisches Journal*, vol. 286, 1892, 158-159.

in the nineteenth century was only a minor issue on a construction site due to the relative inexpensiveness of labour. More important was the speed of distributing and allocating the goods. The mechanisation of the transport system aimed to achieve this goal. The introduction of such interrelated and, by the end of the nineteenth century, common transport systems was never prominently apparent at one single site. Its implementation was a slow and continuous development. As a result, we often tend to neglect the importance of this change and the mechanisation it brought.

Another aspect of mechanisation, in the modern understanding, is the alternation of already existing devices. Many especially smaller devices were optimised in the nineteenth century, primarily as a result of the improved iron production. Optimization does not generally mirror mechanisation,[41] but one possible aspect of optimization was an additional feature of manoeuvrability, which is considered a characteristic of mechanisation. A salient example is the substitution of wood for iron in the construction of barrows, which allowed incorporating an additional way of unloading. This advancement of tool functionality can be defined as mechanisation [fig. 10].

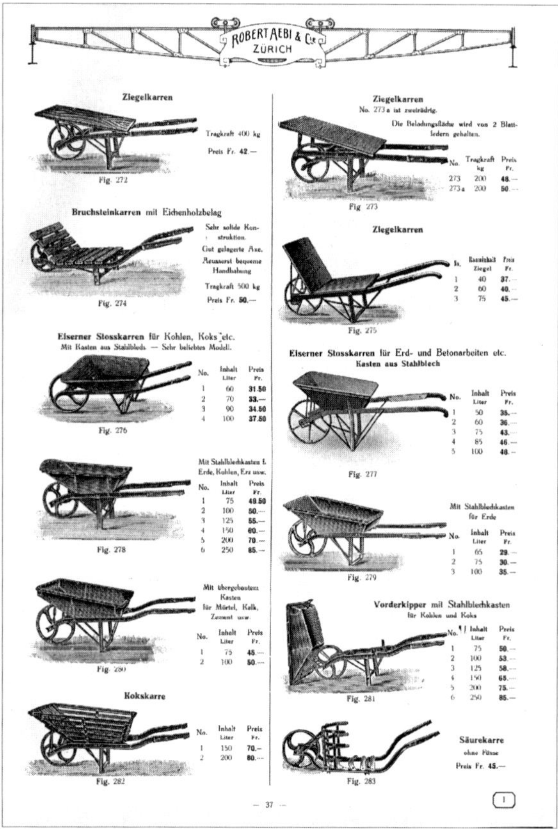

ROBERT AEBI & Cie ZÜRICH

Ziegelkarren

Tragkraft 400 kg

Preis Fr. 42.—

Fig. 272

Bruchsteinkarren mit Eichenholzbelag

Sehr solide Konstruktion.
Gut gelagerte Axe.
Aeusserst bequeme Handhabung

Tragkraft 500 kg

Preis Fr. 50.—

Fig. 274

Elserner Stosskarren für Kohlen, Koks etc.
Mit Kasten aus Stahlblech. — Sehr beliebtes Modell.

No.	Inhalt Liter	Preis Fr.
1	60	31.50
2	70	33.—
3	90	34.50
4	100	37.50

Fig. 276

Mit Stahlblechkasten f. Erde, Kohlen, Erz usw.

No.	Inhalt Liter	Preis Fr.
1	75	49.50
2	100	50.—
3	125	55.—
4	150	60.—
5	200	70.—
6	250	85.—

Fig. 278

Mit übergebautem Kasten für Mörtel, Kalk, Zement usw.

No.	Inhalt Liter	Preis Fr.
1	75	45.—
2	100	50.—

Fig. 280

Kokskarre

No.	Inhalt Liter	Preis Fr.
1	150	70.—
2	200	80.—

Fig. 282

Ziegelkarren
No. 273a ist zweirädrig.

Die Beladungsfläche wird von 2 Blattfedern gehalten.

No.	Tragkraft kg	Preis Fr.
273	200	48.—
273a	200	50.—

Fig. 273

Ziegelkarren

Is.	Raumihalt Ziegel	Preis Fr.
1	40	37.—
2	60	40.—
3	75	45.—

Fig. 275

Elserner Stosskarren für Erd- und Betonarbeiten etc.
Kasten aus Stahlblech

No.	Inhalt Liter	Preis Fr.
1	50	35.—
2	60	36.—
3	75	43.—
4	85	46.—
5	100	48.—

Fig. 277

Mit Stahlblechkasten für Erde

No.	Inhalt Liter	Preis Fr.
1	65	29.—
2	75	30.—
3	100	35.—

Fig. 279

Vorderkipper mit Stahlblechkasten
für Kohlen und Koks

No.	Inhalt Liter	Preis Fr.
1	75	50.—
2	100	53.—
3	125	58.—
4	150	65.—
5	200	75.—
6	250	85.—

Fig. 281

Säurekarre
ohne Füsse
Preis Fr. 45.—

Fig. 283

— 37 —

1

Fig. 10 Diagram with different types of barrows for construction sites, 1912.

41 Paulyini/Trotzsch, *Mechanisierung und Maschinisierung* (fn. 27), 490-495.

Mechanisation, as an additional tool functionality, has also alleviated the burden of labour from workers. One such example is the special stone-pincers that were used to tow dressed stones on vertical lifting devices. Stone-pincers have been used since medieval times[42] and were still in use at around 1900. However, knocking small bores into the stone was a preliminary requirement when using them. This step became redundant when the toggle clamps stone-pincers were invented, which used friction as means of clipping. This extra functionality eased the human work process and thus it can be considered as a tendency toward mechanisation on building sites.

The last two examples focus on small devices. One might, thus, suggest that the increased mechanisation through optimisation of small devices only had a small impact on the construction site as a whole. However, such a perspective fails to take into consideration the usefulness and applicability of the machinery, which in fact strongly differed between different stakeholders.[43] The workers' and foremen's perspectives, whose work was made easier, will have judged differently. They might have seen the trend toward greater mechanisation as a substantial improvement of the work process.

An issue closely connected to mechanisation is the driving power of a process. As previously described, mechanisation, viewed from a nineteenth century perspective, offered the possibility of using machine work (as opposed to work by human labour) for the running of a now mechanised work machine.[44] Thus, the development of small engines, such as gas or electric motors, was seen as a prospect to create an advanced, mechanised workshop and as a mean to 'save' the craftsmen from industrial production. This was a judgement especially articulated by the mechanical engineers, which invented the new engines.[45]

However, motors were rarely in use on the average construction site in the late nineteenth and early twentieth century.[46] This contradicted the contemporary concept of a mechanised workplace and accordingly it seemed that mechanisation did not take place on an average construction site.[47] This viewpoint was basal to research on building machinery until the second half of the twentieth century[48] and can be so even today.

42 Leistikow, Dankwart, "Aufzugsvorrichtungen für Werksteine im mittelalterlichen Baubetrieb: Wolf und Zange", in: *Architectura*, vol. 12, 1982, 20-33.

43 Discussing the different views of different stakeholders in the late nineteenth century: Rauhut, Christoph, "La technique sur le tas. Aspects sociaux de l'utilisation des machines sur les chantiers de construction", in: *Revue d'histoire du XIXe siècle*, vol. 45, 2012, 127-142.

44 Wengenroth, Ulrich, "Motoren für den Kleinbetrieb. Soziale Utopien, technische Entwicklungen und Absatzstrategien bei der Motorisierung des Kleingewerbes im Kaiserreich", in: ibid. (ed.), *Prekäre Selbstständigkeit* (Veröffentlichungen des Instituts für Euorpäische Geschichte Mainz. Abteilung Universalgeschichte, Beiheft 31), Stuttgart 1898, 177-206, 178.

45 Even the famous German engineer Franz Relaueux positioned himself in such a manner: Realeuax, Franz, *Die Maschine in der Arbeiterfrage*, Minden 1885, 20.

46 It is difficult to give precise numbers on the usage, August Vuattolo has tried this for a Swiss case study: Vuattolo, August, *Geschichte des Schweizerischen Bau- und Holzarbeiterverbandes 1873-1953*, vol. 1, Zürich 1953, 87-92.

47 E.g.: Schlieder, Ernst, *Die Maschine im Baugewerbe. Ihre Verwendungsmöglichkeiten und ihre betriebswirtschaftlichen Nutzungsgrenzen*, Dissertation Jena 1928/29.

48 E.g.: Garbotz, Georg, "Baumaschinen einst und jetzt. 100 Jahre Baumaschinen im Spiegel der

However, the perception of the actual usefulness of small motors for a workshop did change in the second half of the twentieth century. Through the work of people such as Karl Heinrich Kaufhold, Ulrich Wengenroth or Helmut Lackner, we now see the differentiated impacts of small engines.[49] One newly discussed aspect was the facilities of the craftsmen and the possibility and necessity of using machine work as a power source.[50] The question of this possibility seems especially important in regards to construction sites, since machine power was a scarcity on many sites.

Some possible problems were that motors needed a stable base, something that, for example, was not an option on a scaffolding; motors were also not particularly mobile. Moreover, because rather steady and slow motions are used when lifiting, such motors always required complex gear ratio devices. Furthermore, the necessity was only sporadic: In many cases building equipment was only used in rather short cycles with longer breaks, the cost of a motor did not offset their demand.

Given the absence of such problems and limitations, one did use motors, for example, to run the travelling cranes.[51] So motors were actually used in cases where it was possible and necessary. Hence, one might conclude that parts of the construction sites were mechanised by the introduction of driving power.

All the examples show yet again that the sites we choose to look at and the way we discuss them are highly critical factors shaping the outcome of our results. Especially, since aspects of mechanisation are not necessarily aspects of a particular site, but are, rather, overarching aspects, which can only be discerned when looking closely at a number of sites. Doing so, one might discover the potential of re-assessing the concept of mechanisation and its use in the history of construction.

3. OUTLOOK

To summarize is to ask: What kind of construction sites should we look at when we set out to critically re-define modern discursive concepts of transformation for contemporary use? Or to be more precise: If we hope to better define concepts such as rationalisation, mechanisation, professionalization, standardisation, economisation and such, as they relate to the realm of construction, and the construction site in particular, what kind of sites should we take into account, and what kind of sites should we dismiss? The following answer is deliberately provocative, but it is intended to clarify the many limitations and pitfalls when attempting an answer:

Bauwirtschafts-Entwicklung", in: *Baumaschine und Bautechnik* vol. 21, 1974, 333-346, 375-388, vol. 22, 1975, 21-32, 79-93, 153-168, 235-254.

49 E.g.: Kaufhold, Karl Heinrich, "Die maschinelle Ausstattung des deutschen Kleingewerbes zu Beginn des 20. Jahrhunderts", in: Bringeus, Nils et al. (ed.), *Wandel der Volkskultur in Europa*, Münster 1988, vol. 2, 835-858; Wengenroth, "Motoren für den Kleinbetrieb" (fn. 44); Lackner, Helmut, "Der Elektromotor als Retter des Handwerks. Mythos oder Realität", in: Plitzner, Klaus (ed.), *Elektrizität in der Geistesgeschichte*, Bassum 1998, 155-168.

50 Kaufhold, "Die maschinelle Ausstattung" (fn. 49).

51 Drews, K., "Entwicklung und gegenwärtiger Stand der modernen Hebezeugtechnik. Die Zeit der zielbewußten Versuche, 1890 bis 1896", in: *Dinglers Polytechnisches Journal* vol. 323, 1908, 17-20.

It seems misleading to solely consider a set of construction sites in which these thematic concepts were visibly or prominently rebuilt, reshaped or reimaged and that were labelled at the time as representatives of the concepts. By solely considering these, we only confirm our assumptions while subtler, hidden aspects are likely to be left unnoticed.

It is therefore necessary to look beyond those sites that were visible and propagated as showcases of modernity through their supposed or actual high level of rationalisation or mechanisation. Accordingly, a history of construction sites also needs to focus on projects that did not fire public and professional imagination. If we consider the last quarter of the nineteenth century, we should not only look at the Eiffel Tower or the Gotthard Tunnel, but rather focus on the broad (and widely unnoticed) building activity, which was primarily housing, but also commercial buildings like offices and state buildings.

In discussing the particular cases of rationalisation and mechanisation above, I have tried to illustrate how it is very likely that different or even new material would be discovered, if we were to focus on this broader set of buildings and uncover the circumstances of their construction sites. Using this material we could define broader and, concurrently, more precise and specific concepts. Or as David Edgerton phrases, when discussing similar thoughts in relation to the history of technology in his seminal work, The Shock of the Old: A "new history" might come to be written, one that "will be surprisingly different".[52]

LIST OF FIGURES

Fig. 1: Deutsches Technikmuseum, AEG-Sammlung.

Fig. 2: Photo by Robert Breitinger 1886, Zentralbibliothek Zürich, Graphische Sammlung/Fotoarchiv, Nachlass Robert Breitinger.

Fig. 3: *The Illustrated London News*, 16 November 1850.

Fig. 4: Gropius, Walter, *bauhausbauten Dessau* (*Bauhausbücher*, 12), München 1930, 171.

Fig. 5: Photo 1899, Baugeschichtliches Archiv der Stadt Zürich.

Fig. 6: Photo by Robert Breitinger April 1900, Zentralbibliothek Zürich, Graphische Sammlung/Fotoarchiv, Nachlass Robert Breitinger.

Fig. 7: Engraving by D. Lancelot, in: *L'Exposition Universelle de 1867 illustrée: publication internationale autorisée par la Commission Impériale*, vol. 1, Paris 1867, p. 116 (Zentralbibliothek Zürich, Alte Drucke).

Fig. 8: *Beton u. Eisen*, vol. 12, 1913, 61.

Fig. 9: Photo by Robert Breitinger 1899, ETH Zürich, gta Archiv 22_056_F_BS_21.

Fig. 10: Robert Aebi & Co., *Werkzeuge und Maschinen für den Eisenbahn-, Strassen-, Hoch- und Tiefbau sowie verwandte Industrien, Ausgabe 1912/13*, part 1, Zürich 1912, 37.

52 Edgerton, David, *The Shock of the Old. Technology and global history since 1900*, London 2006, xi.

LIST OF CONTRIBUTORS

DIPL.-KULT. MARIUS BOETTCHER

is a documentary filmmaker and researcher for Cultural Studies with a focus on Film Studies. After graduating in Media Design as well as in Media Culture at the Bauhaus-University Weimar, he was a Junior Fellow at the International Research Institute for Cultural Technologies and Media Philosophy (IKKM) from 2010 to 2012. His publications include *Das Kinoheft* (co-editor) and *Wörterbuch kinematografischer Objekte* (co-editor). His research interests encompass the modernity in the films of the GDR, the aesthetics of the DEFA-Realism as well as moving images of architecture and its philosophy.

DR. MARIE COLLIER

completed her PhD at The Courtauld Institute of Art in 2015 where she now works as Visiting Lecturer teaching courses on Modernist and Postmodernist architectural history. She is Research Assistant at the Victoria and Albert Museum working on the exhibition "Engineering the World: Ove Arup and the Philosophy of Total Design". From 2011-2015 she was an editor of *Immediations*, the graduate research journal of The Courtauld Institute. In 2011 she contributed to the catalogue *Building the Revolution: Soviet Art and Architecture 1915-1935* at the Royal Academy of Arts London.

DR. SÖREN FISCHER

is a German Art Historian and Curator. 2001-2007 he studied Art History, Classical Archeology and German Literature at the Universities of Münster, Perugia (Italy) and Mainz. In 2007 Master's degree. 2009-2012 he was a research associate at the Institute of Art History in Mainz. 2011 PhD, Doctorate *with Distinction*. 2012-2014 he completed a research traineeship at the Staatliche Kunstsammlungen Dresden. Since August 2014 he is curator of the Museum of Sacred Art in Kamenz/Saxony (Klosterkirche und Sakralmuseum St. Annen). Research and exhibition priorities: Italian Renaissance, Art in the GDR, Contemporary Art. Selected publications: *Das Landschaftsbild als gerahmter Ausblick in den venezianischen Villen des 16. Jahrhunderts: Sustris, Padovano, Veronese, Palladio und die illusionistische Landschaftsmalerei*, Petersberg 2014; *Kreuzigungen. Meisterschüler in Dialogen mit Beckmann, Corinth, Dali, Slevogt und Hrdlicka. Mit neuen Arbeiten der Meisterschüler Kristina Berndt, Caroline Günther, Michael Klipphahn und Winnie Seifert*, Exhibition Catalogue, Kamenz 2015.

DR. EIKE-CHRISTIAN HEINE

graduated 2006 with a M.A. in Politics, Philosophy and Modern History at the Technical University Braunschweig. In 2013 he finished his dissertation about the history of the Kiel. Since 2012 he is "Wissenschaftlicher Mitarbeiter" at the Department for the History of the Impact of Technology at Stuttgart University. His research interests are mostly from the field of the cultural studies and the history of technology as well as knowledge, e.g. the construction of transport infrastructure, environmental history, the history of archaeology, the history of popular culture, or topics from Scandinavian history. His recent publications include *Vom großen Graben. Die Geschichte des Nord-Ostsee-Kanals* (2015); "Connect and Divide. On the History of the Kiel Canal", in: *Journal of Transport History* 35 (2014), S. 200-219.

LILIANA IUGA, B.A.

is doctoral student in Comparative History at the Department of History, Central European University, Budapest (Hungary). She graduated with a B.A. in History and Art History from Babeş-Bolyai University, Cluj-Napoca (Romania), and was recently a visiting student at the Centre for Urban History, University of Leicester (United Kingdom). Her research interests include cultural history, the built environment, nationalism and cultural heritage studies. She is currently working on a dissertation on the politics of urban architectural heritage in Romania during state socialism (1945-1977).

PROF. PATRICK LEITNER

is an architectural and urban designer, teacher, and researcher. He runs his own architectural practice in Paris and holds a position of Associate Professor at the Paris-La Villette School of Architecture where he is also a member of the AHTTEP research group.

In 1998 he started research on urban interactions, mainly of Paris and New York, leading in 2009 to a PhD at the Université Paris-VIII. From the dissertation stems his present research on the perception, discourse, and relationship between cities regarding their built form in reality and mind, for which he uses the concept of Society of Cities. He has published and presented papers on this and other topics at international conferences. Recent publications include *Un ménage urbain à trois, ou l'ambition mondiale de New York* published in Paris Londres (eds. Cohen and Arnold, 2015) and "Paper Experiences: New York's Skyscrapers in the French Public Eye, 1898-1912" to be published in the open access journal *Architectural Histories* (2015, under review).

DR. FILIPPO MENGA

is a Marie Sklodowska-Curie Post-Doctoral Fellow at The University of Manchester, where he is carrying out a study on dams and nation-building through case-studies from Ethiopia and Tajikistan. He holds a Ph.D. in International Relations awarded by the University of Cagliari and he has been visiting researcher at Tallinn University, the University of St Andrews and King's College London. His works have recently appeared in the journals *Nationalities Papers* and *Water Policy*.

DR. CHRISTOPH RAUHUT

is senior research assistant at the Institute of Historic Building Research and Conservation at the ETH Zurich. He studied architecture and received a PhD in 2014. His thesis is entitled: *Die Praxis der Baustelle um 1900: Das Zürcher Stadthaus Fraumünsteramt.* Christoph Rauhut has published extensively, amongst others he is co-editor of *Construction Techniques in the Age of Historicism* (Munich: Hirmer, 2012) and co-author of the to-be-published book *Versuch über die polytechnische Bauwissenschaft.* He is also board member of the Gesellschaft für Bautechnikgeschichte.

IPEK ŞEN, M.A.

is a research and teaching assistant at the Department of Urban and Regional Planning at Istanbul Technical University (ITU), where she is doing her PhD research about location-aware mobile technologies and perception of urban space. After graduating from Mimar Sinan Fine Arts University Urban and Regional Planning, she completed her M.A at Istanbul Bilgi University Cultural Studies Programme. Following her graduation she worked at various Planning and Architecture Firms in Istanbul for several years before starting her PhD at Istanbul Technical University. Currently she is working as a visiting researcher at Cardiff University.

ŞEBNEM ŞOHER, M.SC.

is currently working as a research and teaching assistant at the Department of Architecture at Istanbul Technical University (ITU), where she is continuing her Ph.D studies on urban history and local modernisms. After her B.Arch. degree also from ITU, she completed her M.Sc. in Istanbul Bilgi University Architectural Design Programme. She has contributed various architecture journals based in Turkey, wrote reviews and joined non-profit organisation imkanmekan that works on participatory space-making processes and co-edited *imkanmekan: Small Scale Urban Interventions in Public Space.* Most recently, in 2014, the article "A Section of Urban Memories" that she co-authored with Idil Erkol has been published in *Places of Memory*, official publication of Venice Architecture Biennial Turkish Pavilion.

DR.-ING. CLEMES VOIGTS

is a historian of architecture and engineering specialised in historical building techniques. After his degree in architecture he worked as a scientific assistant and lecturer at the Chair for History of Architecture, Building Archaeology, and Heritage Conservation at Technische Universität München, where he also received his doctoral degree. In 2011 he was awarded a two-year research fellowship at the German Archaeological Institute in Rome. Since 2013 Clemens Voigts teaches as an assistant professor at Universität der Bundeswehr München. Currently he is investigating Late Gothic vault construction in Germany.